High Performance Computing on Vector Systems 2009

Michael Resch · Sabine Roller · Katharina Benkert ·
Martin Galle · Wolfgang Bez · Hiroaki Kobayashi
Editors

High Performance Computing on Vector Systems 2009

Springer

Michael Resch
Sabine Roller
Katharina Benkert
Höchstleistungsrechenzentrum
Stuttgart (HLRS)
Universität Stuttgart
Nobelstraße 19
70569 Stuttgart
Germany
resch@hlrs.de
roller@hlrs.de
benkert@hlrs.de

Martin Galle
Wolfgang Bez
NEC High Performance
Computing Europe GmbH
Prinzenallee 11
40459 Düsseldorf
Germany
mgalle@hpce.nec.com
wbez@hpce.nec.com

Hiroaki Kobayashi
Cyberscience Center
Tohoku University
6-3 Aramaki-Aza-Aoba
Sendai, 980-8578
Japan
koba@isc.tohoku.ac.jp

Front cover figure: Simulation of flow phenomena around a Formula 1 car. Flow acceleration over the rear wing and under the body, flow through the diffuser and roll up in the wake region. Picture due to Department of Aeorspace Engineering, Tohoku University, Sendai, Japan

ISBN 978-3-642-44094-6
ISBN 978-3-642-03913-3 (eBook)
DOI 10.1007/978-3-642-03913-3
Springer Heidelberg Dordrecht London New York

Mathematics Subject Classification (2000): 68Wxx, 68W10, 68U20, 76-XX, 86A05, 86A10, 70Fxx

Cover design: WMX Design, Heidelberg

Printed on acid-free paper

Springer is part of Springer Science + Business Media (www.springer.com)

Preface

This book covers the results of the Teraflop Workbench, other projects related to High Performance Computing, and the usage of HPC installations at HLRS. The Teraflop Workbench project is a collaboration between the High Performance Computing Center Stuttgart (HLRS) and NEC Deutschland GmbH (NEC-HPCE) to support users in achieving their research goals using High Performance Computing.

The first stage of the Teraflop Workbench project (2004–2008) concentrated on user's applications and their optimization for the former flag ship of HLRS, a 72-node NEC SX-8 installation. During this stage, numerous individual codes, developed and maintained by researchers or commercial organizations, have been analyzed and optimized. Within the project, several of the codes have shown the ability to outreach the TFlop/s threshold of sustained performance. This created the possibility for new science and a deeper understanding of the underlying physics.

The second stage of the Teraflop Workbench project (2008–2012) focuses on current and future trends of hardware and software developments. We observe a strong tendency to heterogeneous environments on the hardware level, while at the same time, applications become increasingly heterogeneous by including multi-physics or multi-scale effects. The goal of the current studies of the Teraflop Workbench is to gain insight in the developments of both components. The overall target is to help scientists to run their application in the most efficient and most convenient way on the hardware best suited for their purposes.

The papers in this book demonstrate the different aspects from hard- to software and its applications to various scientific areas. They bring together hardware specialists, supercomputing centers, and users and put their different views on a common topic up to discussions, namely to enable and facilitate leading edge scientific research.

The work in the Teraflop Workbench project gives us insight into the applications and requirements for current and future HPC systems. We observe the emergence of multi-scale and multi-physics applications, the increase in interdisciplinary tasks and the growing tendency to use today's stand-alone application codes as modules in prospective, more complex coupled simulations. At the same time, we notice the current lack of support for those applications. Our goal is to offer to our users an environment that allows them to concentrate on their area of expertise without spending too much time on computer science itself.

We would like to thank all the contributors of this book and the Teraflop Workbench project. We thank especially Prof. Hiroaki Kobayashi for the close collaboration over the past years and are looking forward to intensify our cooperation in the future.

Stuttgart, July 2009

Sabine P. Roller
Michael M. Resch

Contents

I
Petaflop/s Computing

Lessons Learned from 1-Year Experience with SX-9 and Toward the Next Generation Vector Computing

Hiroaki Kobayashi, Ryusuke Egawa, Hiroyuki Takizawa, Koki Okabe,
Akihiko Musa, Takashi Soga, Yoko Isobe

Abstract This paper discusses the performance of the latest vector-parallel supercomputer system SX-9. First, we report HPC challenge benchmark results of the SX-9 system, in which it achieves 19 top one scores in the 28 tests. In addition, we also discuss some tuning techniques for SX-9. We present case study analysis, in which the effects of code tuning for ADB, a newly introduced, software-controllable on chip cache of SX-9, on the sustained performance is examined by using science and engineering applications. Finally, we present our on-going research work on design of a multi-vector core processor toward the next generation vector computing. Through the experimental results, we confirm that vector caching plays an important role to extract the potential of the multi-vector core processor.

1 Introduction

This paper discusses the performance of the latest vector-parallel supercomputer system SX-9 that has been installed at Tohoku University in March 2008. First, we report HPC challenge benchmark results of the SX-9 system. Our SX-9 system achieves 19 top one scores out of 25 tests regarding sustained flop/s, memory and

---

Hiroaki Kobayashi, Ryusuke Egawa, Hiroyuki Takizawa, Koki Okabe
Tohoku University, Sendai 980-8578, Japan, e-mail: koba@isc.tohoku.ac.jp,
egawa@isc.tohoku.ac.jp, tacky@isc.tohoku.ac.jp, okabe@isc.tohoku.ac.jp

Akihiko Musa
Tohoku University, Sendai 980-8578, Japan/NEC Corporation, Tokyo 108-8001, Japan, e-mail: musa@sc.isc.tohoku.ac.jp

Takashi Soga
Tohoku University, Sendai 980-8578, Japan/NEC System Technologies, Osaka 540-8551, Japan, e-mail: soga@sc.isc.tohoku.ac.jp

Yoko Isobe
NEC Computertehno, Ltd., Kofui 400-0055, Japan, e-mail: y-isobe@pi.jp.nec.com

M. Resch et al. (eds.), *High Performance Computing on Vector Systems 2009*,
DOI 10.1007/978-3-642-03913-3_1 © Springer-Verlag Berlin Heidelberg 2010

network bandwidths of representative kernels. In addition, we also discuss some tuning techniques for SX-9. Although SX-9 is equipped with 102.4 Gflop/s vector processors, which are the world-fastest single processors, its bytes per flop rate (B/F) goes down to 2.5. To cover the limited memory bandwidth, a software-controllerable on-chip cache named ADB is available on the chip, which can provide data to the vector pipes at the rate of 4 B/F, instead of the off-chip memory system. Therefore, more ADB-conscious tuning is inevitable to extract the potential of SX-9. We show case study analysis, in which the effect of code tuning for ADB on the sustained performance is examined by using practical science and engineering applications. Through the discussion, we also clarify conflicting effects between ADB tuning and loop unrolling, the latter is one of traditional tuning techniques for conventual vector systems. Finally, we present our on-going research work on design of a multi-vector core processor toward the next generation vector computing. We present an architecture that consists of 4 vector cores sharing an on-chip software-controllable cache, and its early evaluation by using application codes. Experimental results suggest that vector caching plays an important role to extract the potential of the multi-vector core processor.

2 SX-9 System Overview

Figure 1 shows the organization of our SX-9 system. The SX-9 system consists of 16 nodes, and each node has 16 vector processors sharing a large memory space of 1 TB for SMP. The 16 nodes are interconnected via a custom designed high-bandwidth crossbar network named IXS (Interconnect Crossbar Switch) at 256 GB/s

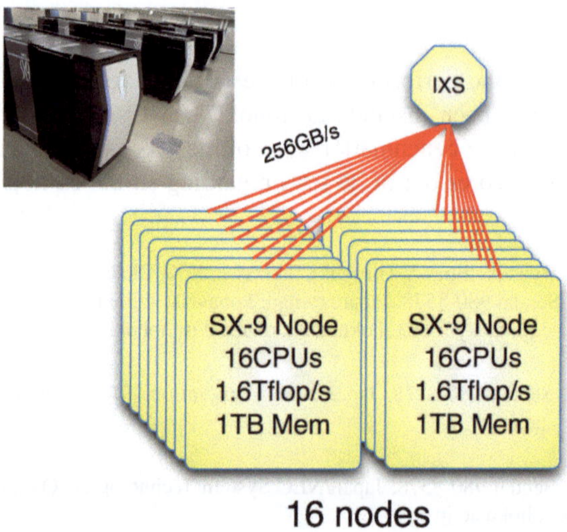

Fig. 1 SX-9 system at Tohoku University

(bi-directonal). As the each processor has a peak performance of 102.4 Gflop/s, the total performance of the system reaches 26.2 Tflop/s.

Figure 2 shows the architecture of the vector processor. The vector processor of SX-9 is designed in the 65 nm technology and its operation frequency is 3.2 GHz. The SX-9 processor also employs a 256 KB on-chip cache, which is newly designed for the SX vector processor named ADB (Assignable Data Buffer) to assist the vector load-store unit. By keeping vector data with the locality on ADB, a 4 B/F (bytes per flop) rate is guaranteed to effectively drive plenty of vector pipes on the chip. Table 1 summarizes the performance of the SX-9 system. The balanced combination of high performance vector CPUs and a larger shared memory, which is connected together at higher memory bandwidth, provides user-friendly high-performance computing environments, in which users enjoy the potential of the system with a less efforts.

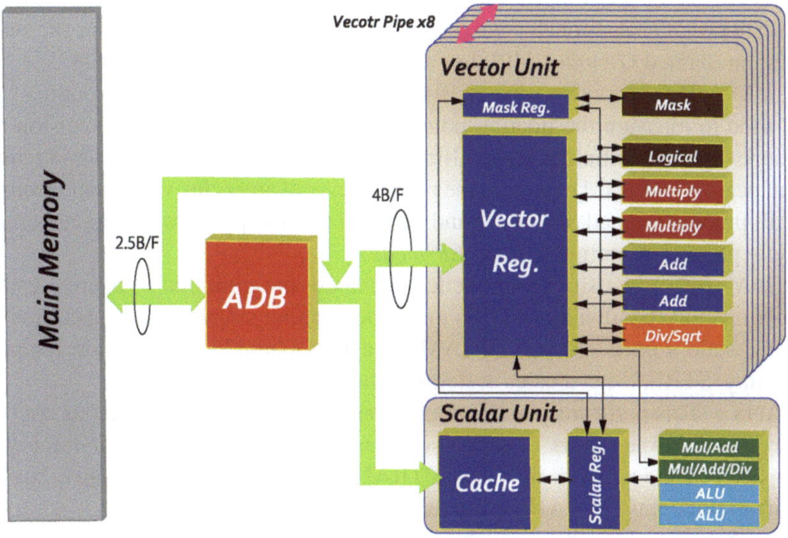

Fig. 2 SX-9 processor architecture

Table 1 Performance Specification of SX-9

per CPU	Freq.	3.2 GHz
	Vector Pipes Performance	102.4 Gflop/s
	Memory Bandwidth	256 GB/s
per SMP node	Vector Pipes Performance	1.6 Tflop/s
	Memory Capacity	1 TB
	Memory Bandwidth	4 TB/s
	Memory Banks	32 K
	IXS Bandwidth	256 GB/s
per System	Total Performance	26.2 Tflop/s
	Total Memory capacity	16 TB

3 HPC Challenge Benchmark Results

To evaluate the potential of the SX-9 system, we carried out it performance evaluation by using the HPC Challenge benchmark. The HPC challenge benchmark suite [1] has been designed and developed by the DARPA HPCS (High Productivity Computing Systems) program to evaluate high-end computing systems from the wide variety of viewpoints, not only HPL performance in flop/s, which is used for TOP500 ranking, but also memory performance that seriously affects the sustained system performance in the execution of real applications. To this end, the HPC challenge benchmark focuses on the memory bandwidth evaluation by using several kernels that have different degree of the spatial and temporal locality of memory references. The suite consists of basically seven categories, and each category includes several tests in the different modes (total 28 tests):

HPL This is the LINPACK TPP benchmark that measures the floating point rate in Tflop/s of execution for solving a linear system of equations on the entire system (named *G-mode*). HPL has a high spatial and temporal locality.

DGEMM This is a subroutine of BLAS (Basic Linear Algebra Subroutines) and is used to measure the floating point rate of execution of double precision real matrix-matrix multiplication in Gflop/s. DGEMM has a high spatial and temporal locality. DGEMM evaluates the performance of an exclusively running single-MPI process (named *SN-mode*) and per-MPI process performance in embarrassingly parallel execution (named *EP-mode*).

STREAM This is a simple synthetic benchmark program that measures sustainable memory bandwidth in GB/s under performing simple vector operations of copy, scale, sum and triad. STREAM has a high spatial locality, but low temporal locality. STREAM evaluates systems in the SN-mode and EP-mode.

PTRANS This performs parallel matrix transpose through simultaneous communications between pairs of processors, and measures the total communication capacity of the network in GB/s. PTRANS has a high spatial locality, but low temporal locality. PTRANS evaluates systems in the G-mode.

RandomAccess This measures the rate of integer random updates of memory in GUPS (Giga Updates Per Second). As RandomAccess generates highly irregular memory accesses, it has a low spatial and temporal locality. RandomAccess evaluates systems in the G, SN and EP modes.

FFTE This is a kernel program to measure the floating-point rate of execution of double precision complex one-dimensional Discrete Fourier Transform. FFTE has a low spatial locality, but high temporal locality. FFTE evaluates systems in the G, SN and EP modes.

Communication Bandwidth and Latency This measures latency and bandwidth of a number of simultaneous communication patterns (PingPong, natural-ordered rings and random-ordered ring). To measure the bandwidth and latency, 2 M bytes and 8 bytes of data are used, respectively.

Figure 3 is a radar chart showing the results of SX-9 on the 28 tests of the HPC challenge benchmark with some representative high end computing systems such as IBM BG/P and CRAY TX3/4/5. System configurations discussed in this section are summarized in Table 2. The outermost circle is the position for the system with the best performance on each test, and other systems are plotted according to their performance normalized by the best one in the range between 0 (worst) at the center and 100% (best) on the outermost circle. The detailed results are available on the HPC Challenge benchmark web site at http://icl.cs.utk.edu/hpcc/hpcc_results_all.cgi.

As the radar chart shows, the SX-9 system achieves total 19 top one scores in 28 tests: 10, 5, and 4 top ones on memory bandwidth, network bandwidth and flop/s tests, respectively. In the following, we take a close look at some of individual results.

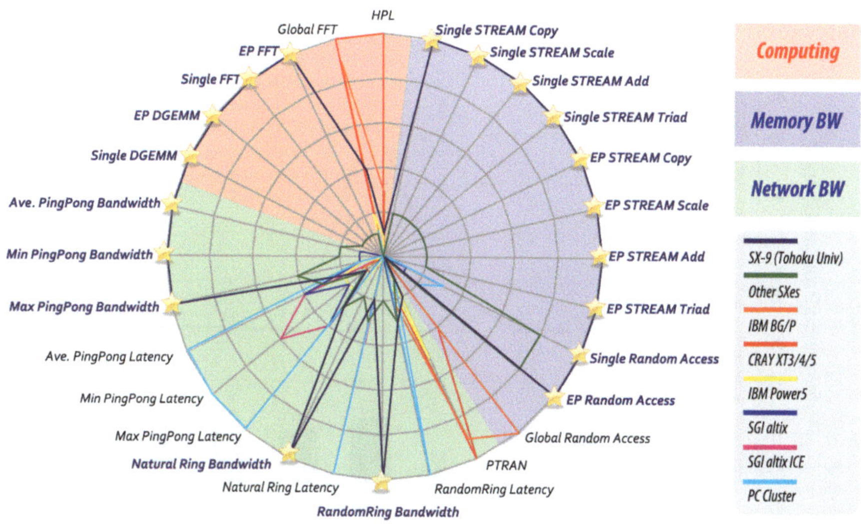

Fig. 3 HPC challenge benchmark results

Figure 4 shows the HPL results. As the HPL test evaluates the entire system performance in G-mode, the absolute performance of our 26.2 Tflop/s SX-9 system on this test is low compared with other peta and sub-peta level systems such as CRAY XT-5 and BG/P. However, when looking at the performance per process, the SX-9 system shows the significant performance owing to its world-fastest single vector processor with 102.4 Gflop/s. The efficiency of the SX-9 system on this test remains 77%, however, according to the recent technical report about HPL performance [2], SX-9 of Earth Simulator Center achieves 93.4%, top one efficiency in the LINPACK performance at TOP 500 ranking.

In the cases of memory-related tests such as STREAM tests, the SX-9 system achieves outstanding sustained performance compared with those of the scalar systems. A test in the STREAM category reads data from memory, performs some operations, and the writes back data to the memory. The performance is measured

Table 2 List of evaluated systems

System name	manufacture	Processor Type	Freq.	Cores	MPI Procs	Threads	Peak [TF]	Interconnect	Network BW
SX-9	NEC	SX-9	3.2 GHz	256	256	1	26.2	IXS	128 GB/s
SX-9(SMP)	NEC	SX-9	3.2 GHz	32	2	16	3.2	IXS	128 GB/s
SX-8	NEC	SX-8	2 GHz	40	40	1	0.64	IXS	16 GB/s
SX-8(SMP)	NEC	SX-8	2 GHz	40	5	8	0.64	IXS	16 GB/s
SX-7	NEC	SX-7	0.552 GHz	32	32	1	0.28256	non	non
SX-7(SMP)	NEC	SX-7	0.552 GHz	32	2	16	0.28256	non	non
Idataplex	IBM-Serviware	Xeon X5472	3.0 GHz	1088	1,088	1	13.5	Infiniband	2 GB/s
BL2x220	HP	Xeon E5450	3.0 GHz	256	256	1	3.072	Infiniband	2 GB/s
SC5832	SiCortex	SiCortex Ice9	0.7 GHz	5760	5,760	1	8.064	Custom	6 GB/s
Blue Gene/P	IBM	PowePC450	0.85 GHz	131,072	131,072	1	557	Torus	425 MB/s
Blue Gene/P (SMP)	IBM	PowePC450	0.85 GHz	131,072	32,768	4	557	Torus	425 MB/s
XT5	CRAY	AMD Opteron	2.3 GHz	149,058	74,529	2	1,381.62	Seastar	9.6 GB/s
Darwin	ClusterVision/ Dell/QLogic	Xeon 5160	3 GHz	256	256	1	3.072	Infiniband	2 GB/s
Altix 8200EX	SGI	Xeon X5472	3 GHz	1,024	1,024	1	12.288	Infiniband	2 GB/s
Intel Endeavor cluster	Intel	Xeon 5160	3 GHz	1,024	1,024	1	11.4688	Infiniband	2 GB/s

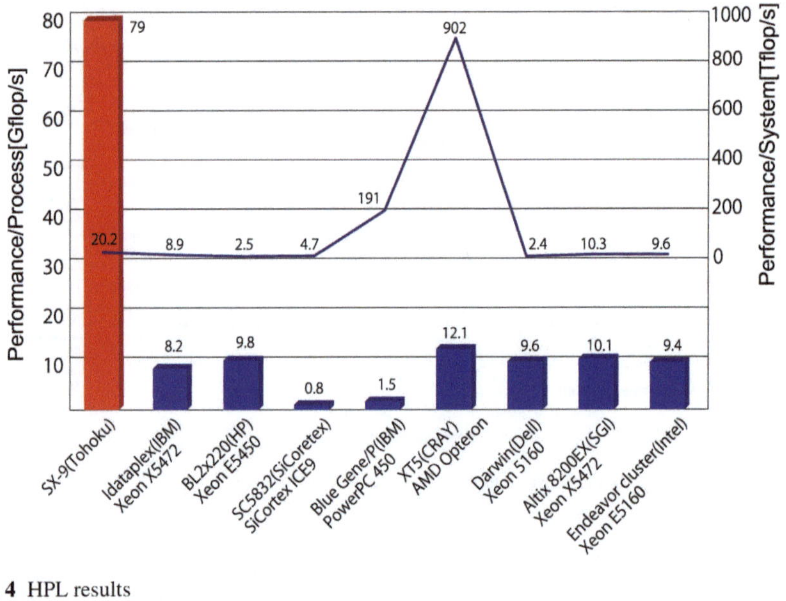

Fig. 4 HPL results

as the effective memory bandwidth of the systems in GB/s. Figure 5 shows the results of two types of runs on the SX-9 system: one is a flat MPI run with 256 MPI processes on 256 CPUs, and the other is a hybrid run with 2 MPI processes on 2 nodes, each of which is a 16-CPU SMP on a single node. The latter is available only on SX-9 with a larger SMP node with 16 CPUs, capable of 1.6 Tflop/s with a 4 TB/s memory bandwidth. As the graph shows, SX-9 shows a significant performance compared with scalar systems. Cray XT5 with hybrid configuration shows the best performance among scalar systems, however, the SX-9 performance is still 20 times higher even in the case of the flat MPI mode. In the case of hybrid configuration of 2 16-SMP threads, the performance is boosted by a factor of 13, now reaches 2.7 TB/s. The performance in the stream test completely reflects the feature of the vector system regarding its high-memory bandwidth.

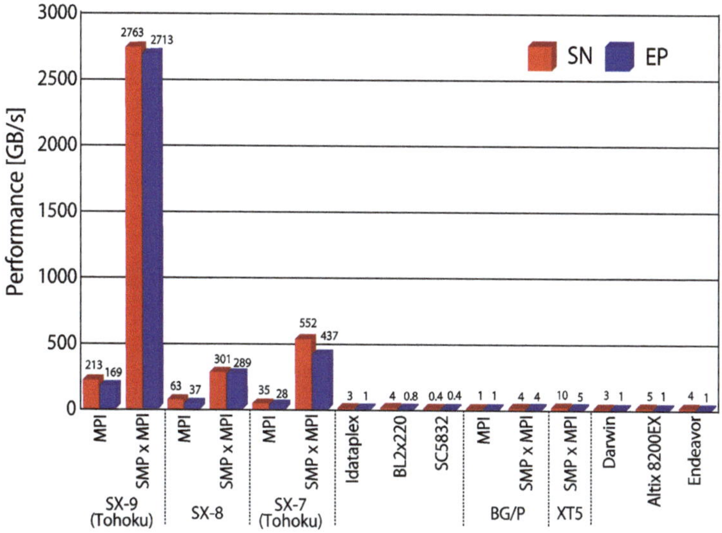

Fig. 5 STREAM results

Figure 6 shows the results on SN and EP tests in the Random Access category measured in the unit of giga update per second. The tests in this category check performance in moving individual data rather than large arrays of data. Moving individual data quickly and well means that the computer can handle chaotic situations efficiently. As these tests enforce highly-irregular memory accesses on the systems, the cache memory of the scalar systems does not work efficiently. On the other hand, higher memory bandwidth of vector systems is very effective for applications even with random memory accesses. The SX-9 system shows 20-times larger performance than CRAY XT5 that again shows the best performance among the scalar system in this test. However, we also point out the performance of SX-9 does not scale well compared to SX-8 and SX-7. This is because of its lower B/F rate and longer memory latency compared with the other SX systems.

Figure 7 shows the performance of Global FFT test in G-mode. As this test tends to highly reflect the size of the systems, larger systems with higher peak performance are advantageous to achieve higher performance. The best performance of the G-FFT is achieved by Cray XT5 and second one is IBM blue gene, they are peta-flop/s level machines and marked around 5 Tflop/s performance in this test. Our SX-9 is placed at the fourth with a sustained performance of 2.38 Tflop/s. But when looking at the single thread performance, our SX-9 shows the significant performance, which is 100-times higher than Cray XT5 regarding per-process performance as shown in this figure.

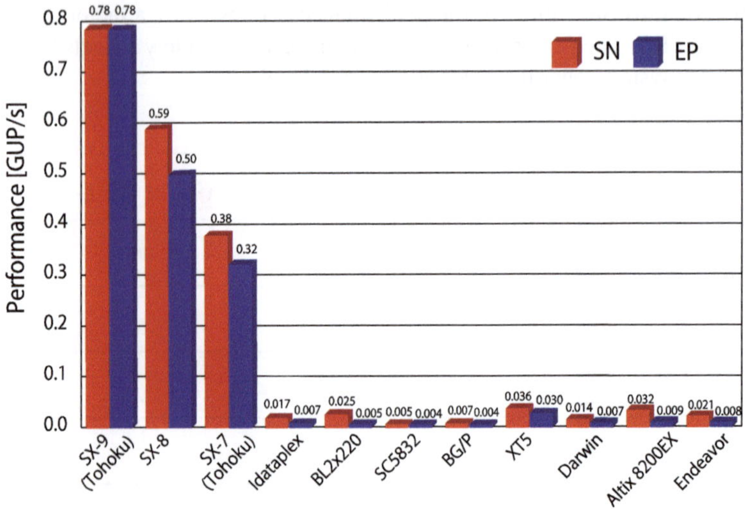

Fig. 6 Random access test results

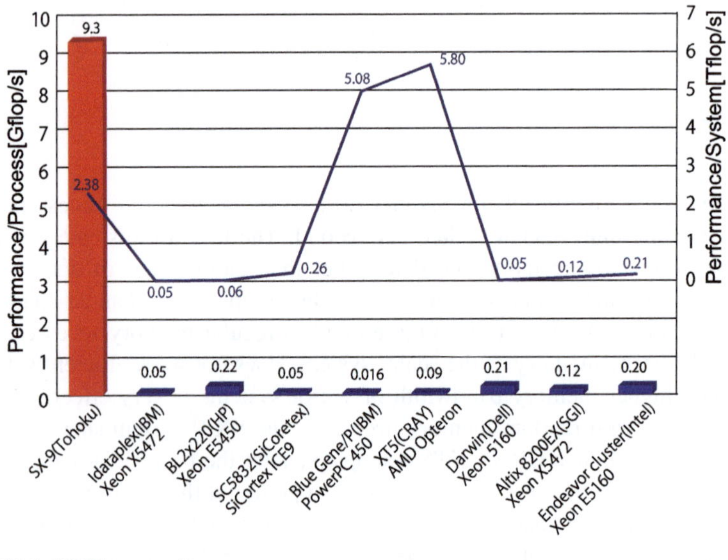

Fig. 7 Global FFT test results

Table 3 summarizes the details of performance comparison. As this table shows, top three systems are peta and sub peta level systems, and our SX-9 is at the fourth position behind them, but regarding the efficiency of the sustained performance to the peak performance, Cray XT5 at the top 1 position that has 52 times higher peak performance realized by almost 600-time more processor cores shows only 2.5 times higher the performance compared with our SX-9, resulting in 23 times higher efficiency of SX-9.

Table 3 Performance comparison in G-FFT test

RANK	System	Institution	Peak Perf. (Tflop/s)	CPUs (Cores)	G-FFT Results (Tflop/s)	Efficiency
1	Cray XT5	Oak Ridge National Lab.	1381.6	37544 (150176)	5.8	0.40%
2	BlueGene/P	Argonne National Lab.	557	32768 (131072)	5.1	0.90%
3	Red/Storm/XT3	Sandia National Lab.	124.4	12960 (25920)	2.9	2.30%
4	SX-9	Tohoku Univ.	26.2	256 (256)	2.3	9.10%

Figure 8 shows network latencies of individual systems in Random Ring and PingPong tests. There is no significant difference among the systems regarding latency, except for the Idataplex. InfiniBand of Darwin, which is a well-used commodity network, shows the shortest latency in these tests. IXS of SX-9 needs a certain amount of the setup overhead which is larger than that of infini-band because the data should be prearranged for the effective use of its 16 parallel channels, in order to achieve a high network bandwidth of 256 GB/s. As this setup is mainly processed as scalar operations, its leads to a bit longer latency than the commodity networks, even though IXS is the custom network.

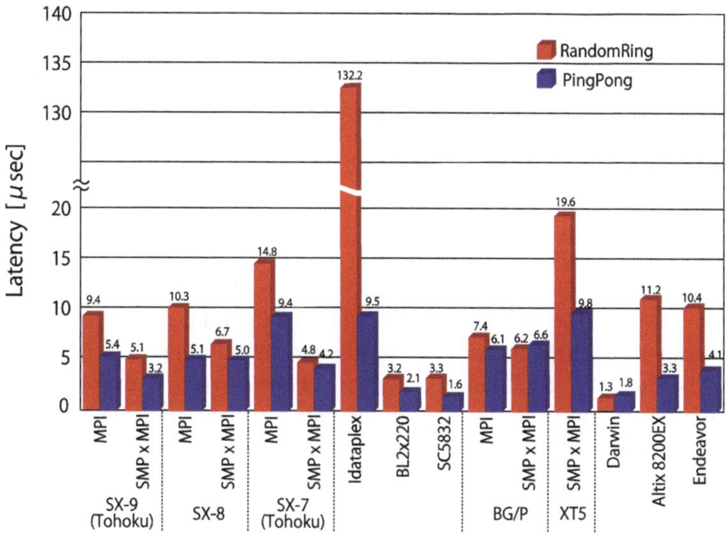

Fig. 8 Network latency test results

In the case of bandwidth tests, IXS outperforms the other network systems. Figure 9 shows sustained bandwidths of individual systems in random ring and pingpong tests. In the random ring, a pair of threads to communicate each other are randomly selected. Therefore, the random ring test gives a high pressure on the network system, and as a result, the bandwidth is limited by inter-node communication performance of the systems. In the hybrid run, as the number of threads that needs the communication is reduced, the load on the network is relaxed, resulting in an improvement in bandwidth per thread. In the PingPong test, sequentially activated ping-pong type communication also relaxes loads over the network, and therefore, the bandwidth is further improved even in the case of the flat MPI run. As the graph shows, SX vector systems achieve quite higher sustained bandwidth compared with the scalar systems. However, we also point out the sustained bandwidth of the SX-9 system remains only 23% of the peak bandwidth, which is 128 GB/s in each direction. This is due to the size of data exchanged between the threads in the tests.

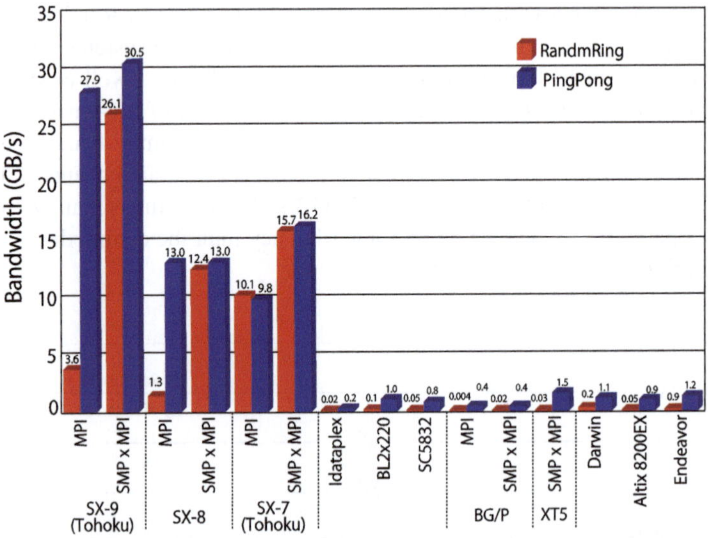

Fig. 9 Network bandwidth test results

Figure 10 shows the sustained performance of the ping-pong test as a function of the size of its payload in MB. The results are slightly higher from those shown in Fig. 9 because the results in Fig. 10 are obtained by the execution of the pure ping-pong kernel only, not obtained by the entire execution of the HPC challenge benchmark suit, which needs an additional overhead. In the HPC challenge benchmark, the data size is specified to 2 MB, and only 23% of the peak bandwidth is obtained. However, if we increase the size to more than 100 MB, 80 to 90% of the peak can be obtained. Therefore, these experimental results suggest that when we develop the applications for SX-9, the data size is a very important factor to achieve higher sustained network bandwidth. In addition, as IXS of SX-9 can handle up to 16 independent data transmission between MPI processes simultaneously, aggregated use of individual MPI communications is also a tuning option to increase the sustained bandwidth of IXS.

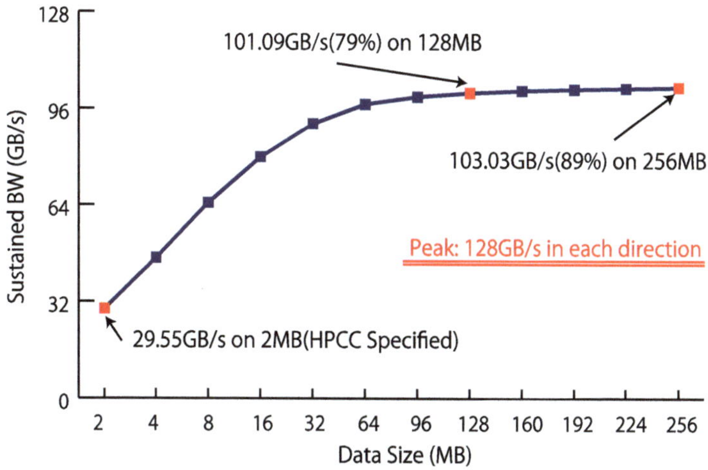

Fig. 10 Data size vs. sustained network bandwidth

4 Case Study Analysis of Memory-Conscious Tuning for SX-9

So far, vector users tend to pay attention to increases in vector lengths only, not to reducing memory operations because of its high memory bandwidth. However, as 4 B/F is no longer available on the SX-9 system, memory-conscious code design becomes important to achieve higher sustained performance. When considering the feature of the SX-9 architecture, we have several options for performance improvement regarding reducing off-chip memory transactions.

The first one is selective caching. As ADB is software-controllable and its size is only 256 KB, we have to effectively use the limited on-chip space to have the benefit of 4 B/F between ADB and the register file. Therefore, we carefully select vector data with the high locality of reference in codes and specify them for caching, instead of all the data cached in the conventional scalar system. Vector data caching is very promising because one vector data reference with the locality gives us a great number of data references with locality at the array element level. Therefore we can expect that hits of vector data on ADB results in a significant gain as a sustained performance improvement.

In addition, cache blocking is also important to increase cache hit rates, in order to avoid the capacity misses. At the same time, we have to be careful to apply cache blocking to vector kernels, because it may decrease vector lengths, and shorter vectors affect the efficiency of vector processing. Therefore, there is a trade-off between cache blocking and vector length depending on the characteristics of individual codes.

Besides, we should also be careful when applying the traditional tuning options to the codes for SX-9. Loop unrolling and loop fusion are typical tuning techniques to enlarge a benefit on vector systems, for example,

1. increases in both arithmetic density and vector length in loop bodies,
2. a decrease in the branch overhead, and
3. an increase in temporal locality of data by removing duplicated vector loads across nested loops.

However, they also increase the possibility of register spills and/or evictions from ADB if its capacity is not enough to hold data in enlarged loop bodies by loop unrolling and loop fusion. Therefore, excessive loop unrolling and/or loop fusion may also lead to a harmful effect on vector systems with limited off-chip memory bandwidths, such as SX-9.

In the following discussion, we examine some tuning techniques for SX-9 by some representative high cost kernels excerpt from real applications.

The first example for ADB optimization is a plasma simulation code, which is a code for the simulation of upper hybrid wave in plasma using Lax-Wendroff method [3]. Figure 11(a) is one of high cost kernels of Plasma. In this loop, array *dvecw* are indirectly accessed through index *ic*. Therefore, this kernel needs gather operations for indirect memory accesses. The memory access latency of a gather operation is exposed in the execution time because the address calculation related to the gather operation cannot be hidden. However, as *dvecw* are defined before the loop, selective caching on ADB by directive *cdir on_adb(dvecw)* works well. By selective caching of *dvecw* with its blocking for the 256 K ADB, all necessary processing on this array is carried out on ADB. As a result, we can reduce the memory access time to one-third, resulting in a 2.3x performance improvement, compared with that of the original code without ADB caching as shown in Fig. 11(b).

Another case is to calculate the sum of multiple-add operations of arrays shown in Fig. 12(a). In the original code, $sum(i, 1, k)$ and $sum(i, 2, k)$ have the locality of reference, and are the best candidate for caching on ADB. However, they need 1.6 MB each to hold the entire data on ADB. Therefore, cache blocking is also needed to fit the data onto ADB. Figure 12(b) is a tuned code with caching on ADB. Here *vlen* is a cache blocking parameter. If *vlen* is equal to or less than 12,800, $sum(i, 1,)$ and $sum(i, 2, k)$ are cacheable on the 256 KB ADB. Figure 12(c) shows the sustained performance as a function of the size of *vlen*. As the size of *vlen* increases, the performance also increases accordingly, and when *vlen* reaches 12,800 , the best performance of 54.7 Gflop/s on a single processor, which is 60% more performance than the original, is obtained. However, further increase gives a harmful effect on the performance. The performance of the original size, $vlen = 100{,}000$, is 37% lower than the best one. Therefore, simply increasing loop size does not lead to the performance improvement on SX-9, when the data is provided from ADB, not from the off-chip banked memory.

```
!cdir on_adb(dvecw), nodep
 do ii = jj, min(jj+lvec-1, iplast)
   kk = ii-jj+1
   aa = xp1(ii)/delx
   ic = int(aa+half)
   raa1 = ic
   dd1 = fact1*(aa+half-raa1)
   dd2 = fact1*(raa1-aa+half)
   dvecw(ic,kk, 1) = dvecw(ic, kk, 1)+dd1*vp1(1,ii)
   dvecw(ic-1, kk, 1) =dvecw(ic-1, kk, 1)+dd2*vp1(1,ii)
   dvecw(ic, kk, 2)=dvecw(ic, kk, 2)+dd1*vp1(2,ii)
   dvecw(ic-1, kk, 2)=dvecw(ic-1, kk, 2)+dd2*vp1(2,ii)
   dvecw(ic, kk, 3)=dvecw(ic, kk, 3)+dd1*vp1(3,ii)
   dvecw(ic-1, kk, 3)=dvecw(ic-1, kk, 3)+dd2*vp1(3,ii)
   dvecw(ic, kk, 4)=dvecw(ic, kk, 4)+dd1
   dvecw(ic-1, kk, 4)=dvecw(ic-1, kk, 4)+dd2
 enddo
```

(a) Kernel Code

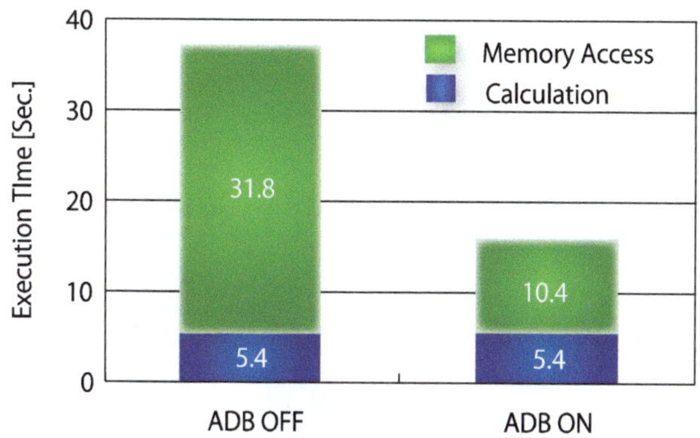

(b)ExecutionTime

Fig. 11 Effect of ADB (plasma)

The next case is a code for differential equations, frequently used in scientific and engineering applications. Figure 13(a) is a kernel of an electro-magnetic field simulation based on the FDTD model for the design of a land mine detector [4]. As data references in the differential equation of the FDTD model have a high locality of reference, this kind of kernels is best candidate for caching on ADB. In this example, arrays H_x, H_y, and H_z have the locality regarding index i and k. For example, in the i-th iteration, if preceding $H\_y(i, j, k)$ vector load brings data into ADB, most of all the data can be provided from ADB in the vector load of $H\_y(i - 1, j, k)$ in the $i + 1$ th iteration, resulting in a high ADB hit rate. Similarly, arrays H_x and H_z also have the locality. For directing selective caching of H_x, H_y, and H_z, directive *cdir ON_ADB(H_x)*, *ON_ADB(H_y)*, and *ON_ADB(H_z)* are needed. Figure 13(b) shows the performance of SX-9 with ADB normalized by that without ADB. As the graph shows, the time for memory accesses can be reduced

```
do k=1,l
  do j=1,m
    do i=1,n
       sum(i,1,k) = sum(i,1,k) + dble(a(j,k))*b(i,1,j)
&          - aimag(a(j,k))*b(i,2,j)
       sum(i,2,k) = sum(i,2,k) + dble(a(j,k))*b(i,2,j)
&          + aimag(a(j,k))*b(i,1,j)
      end do
   end do
end do
```

(a) Original code

```
do k=1,l
  do i2=1,n,vlen
    do j=1,m
!cdir on_adb(sum)
      do i=i2,min(n,i2+vlen)
        sum(i,1,k) = sum(i,1,k) + dble(a(j,k))*b(i,1,j)
&           - aimag(a(j,k))*b(i,2,j)
        sum(i,2,k) = sum(i,2,k) + dble(a(j,k))*b(i,2,j)
&           + aimag(a(j,k))*b(i,1,j)
       end do
     end do
  end do
end do
```

(b) Code with selective caching and blocking

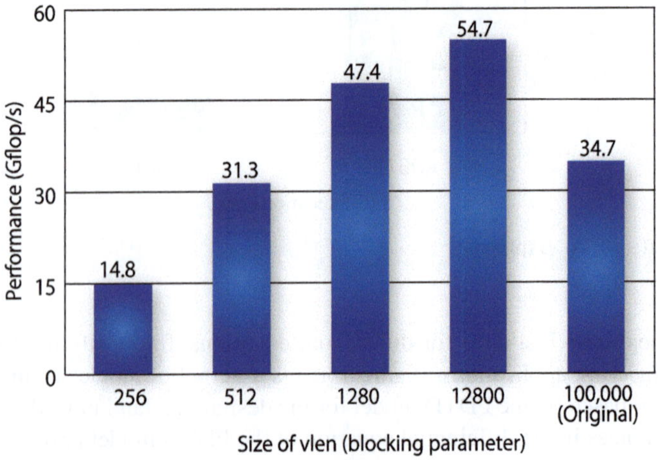

(c) Performance

Fig. 12 Effect of cache blocking on multiply-add kernel

by 34% and a 20% improvement in performance is obtained by selective caching on ADB in this case.

Basically, loop unrolling is an effective tuning option for vector systems, however, there are some cases that excessing unrolling gives a harmful effect on SX-9.

```
DO 10 k=0,Nz
  DO 10 i=0,Nx
!cdir ON_ADB(H_x)
!cdir ON_ADB(H_y)
!cdir ON_ADB(H_z)
    DO 10 j=0,Ny
       E_x(i,j,k) = C_x_a(i,j,k)*E_x(i,j,k)
&       + C_x_b(i,j,k) * (( H_z(i,j,k) -H_z(i,j-1,k))/dy
&       -(H_y(i,j,k) -H_y(i,j,k-1))/dz -E_x_Current(i,j,k))
       E_y(i,j,k) = C_y_a(i,j,k)*E_y(i,j,k)
&       + C_y_b(i,j,k) * ((H_x(i,j,k) -H_x(i,j,k-1))/dz
&       -(H_z(i,j,k) -H_z(i-1,j,k))/dx -E_y_Current(i,j,k))
       E_z(i,j,k) = C_z_a(i,j,k)*E_z(i,j,k)
&       + C_z_b(i,j,k) * ((H_y(i,j,k)-H_y(i-1,j,k) )/dx
&       -(H_x(i,j,k)-H_x(i,j-1,k))/dy -E_z_Current(i,j,k))
10 CONTINUE
```

(a) Kernel Code for Difference Equation

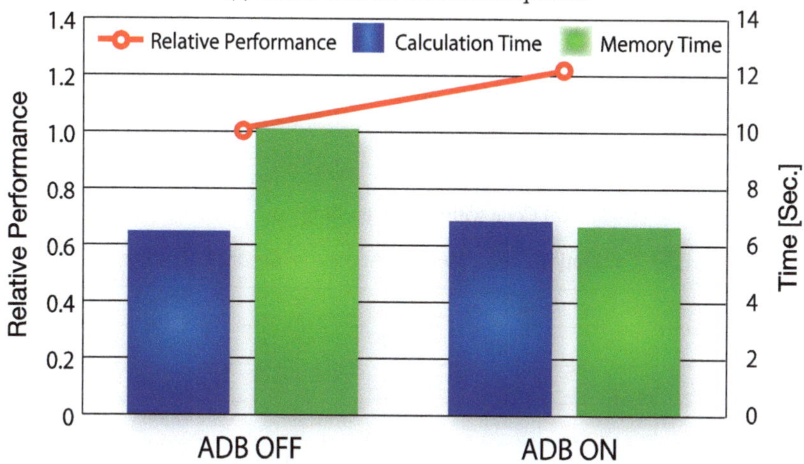

(b) Performance

Fig. 13 Effect of ADB on kernel for difference equation (land mine)

Figure 14(a) is an example of that case, which is one of high cost kernels of an earthquake simulation code [5]. This kernel can be unrolled with outermost loop as shown in Fig. 14(b). Figure 14(c) shows the performance of this kernel as a function of degree of unrolling. As unrolling proceeds, the performance is improved due to a decrease in the number of memory references. However, the performance decreases with further loop unrolling, and in this case, unrolling 8 times shows the best performance. This is a typical example that shows excessive unrolling also increases memory operations due to register spill and evictions of effective data from ADB, although this is not a case for former SX systems with the 4 B/F ratio, which can handle more larger degree of loop unrolling to improve the performance. As a result, the time for memory-related operations on the SX-9 system is gradually exposed in the execution time as the degree of unrolling increases.

```
do km = 1, nd
do jq = 1, nsum_dip
 do iq = 1, nsum_dip
    sum1(iq,km) = sum1(iq,km)+ r(jq,km) * gd(iq,jq,km)
    sum2(iq,km) = sum2(iq,km)+ i(jq,km) * gd(iq,jq,km)
 end do
end do
end do
```

(a) Original code

```
   do km = 1, nd
    do jq = 1, nsum_dip, k
    do iq = 1, nsum_dip
      sum1(iq,km) = sum1(iq,km)+r(jq,km)*gd(iq,jq,km)
&     +r(jq+1,km)* gd(iq,jq+1,km)+r(jq+2,km)* gd(iq,jq+2,km)
&  ...
&     +r(jq+k-1,km)*gd(iq,jq+k-1,km)
      sum2(iq,km) = sum2(iq,km)+ i(jq,km)*gd(iq,jq,km)
&     +i(jq+1,km)*gd(iq,jq+1,km)+ i(jq+2,km)*gd(iq,jq+2,km)
&   ...
&     +i(jq+k-1,km)* gd(iq,jq+k-1,km)
    end do
  end do
 end do
```

(b) Code with Loop Unrolling

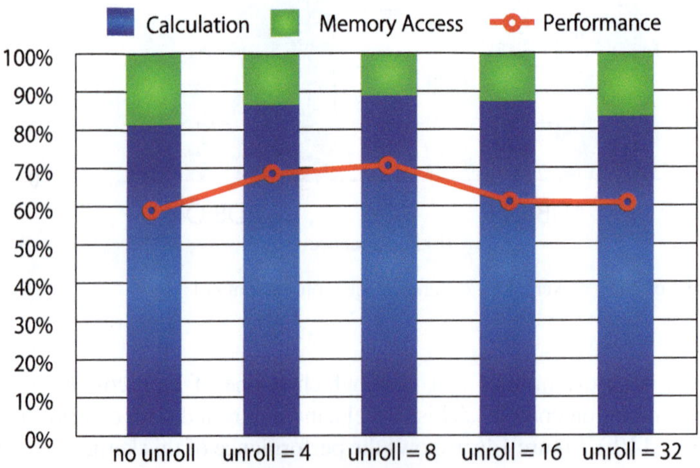

(c) Breakdown of Execution Time

Fig. 14 Effect of loop unrolling on earthquake kernel

5 Multi-Vector Cores Processor Design

Through the discussion on ADB, we confirmed its potential to cover the limited off-chip memory bandwidth of SX-9. We think an on-chip memory system is now becoming a key component toward next generation vector computing even in chip multiprocessing, as commodity multi-core processors now employ a large on-chip cache.

Thanks to advances in circuit integration technologies, chip multiprocessors (CMPs) have become the mainstream in commoditybased scalar processors. Eight-core CMPs are already found in the commercial market. CMP-based vector processors have not been available in the commercial market yet, however, the CMP architecture is also promising for vector processor design, because the number of transistors in a vector processor has been increasing by a factor of eight for the last decade.

We are working on design of a chip multi-vector core processor (CMVP) with a shared cache. Figure 15 shows the CMVP block diagram, which has four vector cores sharing a software-controllable vector cache. The main memory unit is an interleaved memory system and is divided into 32 parts, each of which operates independently and supplies data to the cores. To cover the limited off-chip memory bandwidth, sub-caches of the vector cache are connected to these memory ports; the vector cache consists of 32 sub-caches. Moreover, the vector cores and the sub-caches are interconnected through 32 crossbar switches at 4 B/F. To realize direct accesses to both the off-chip memory and the vector cache simultaneously, a vector-cache bypassing mechanism is also introduced. Holding data on the cache and bypassing are all software-controllable through software directives at the programming level.

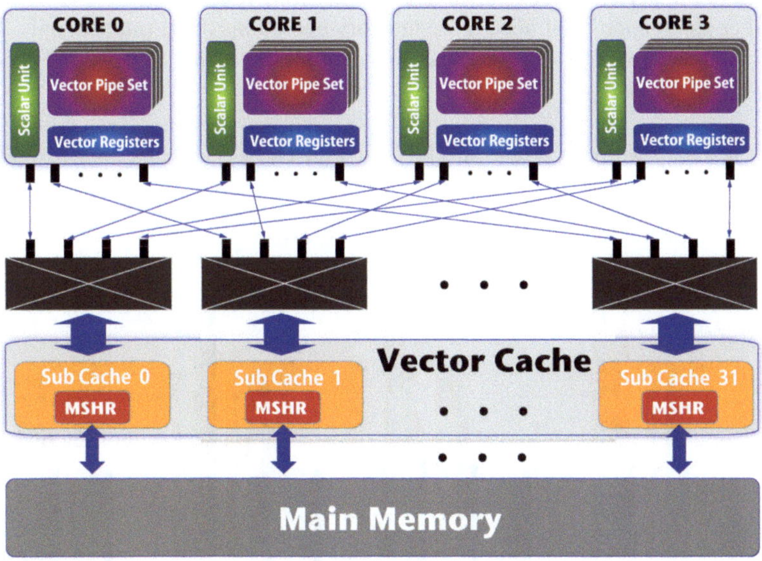

Fig. 15 Multi-vector core processor architecture

The sub-cache is a non-blocking cache and includes a tag array, a data array and a MSHR (Miss Status Handling Registers). The sub-cache employs a set-associative write-through cache with the LRU replacement policy. The line size is 8 bytes; it is the unit for memory accesses. Here, the MSHR holds the information of in-flight load requests: instruction address and memory address of load data. When the memory address of a subsequent load request is equal to that of an in-flight load data, the subsequent load requests are not sent to the main memory and wait for the arrival of data to be brought to the MSHR by the preceding outstanding load request. Then, the subsequent load requests are immediately solved as the in-flight load data arrives at the MSHR. This mechanism can reduce the number of the load request between the vector cache and the main memory.

We evaluate the effect of the vector cache on CMVP by using a cycle accurate simulator we have developed. In the evaluation, we examine the effect of B/F rates of the CMVP, changed from 1 to 4, on the sustained performance. For the evaluation, we use four scientific applications: Jacobi (a linear equation solver), Land Mine [4], Antenna [6], and Earthquake [5].

Figure 16 shows scalabilities of CMVPs with different B/F rates evaluated by using the four applications. For comparison, CMVPs without the cache are also examined. The 4 B/F system alway shows the scalable performance with up to 4 cores in all the applications even without the cache. However, when the B/F rate is limited to 2 or 1, the scalability declines with the increasing number of cores. In such cases, the cache can boost the performances. For example, in Antenna, which is a computation-intensive application, the cache can assist the 1 B/F and 2 B/F systems to achieve the performance comparable to that of the 4 B/F system. Meanwhile, in the other applications that need more memory operations than Antenna, only 2 B/F system can achieve the scalable performance with the cache. In the 2 B/F system,

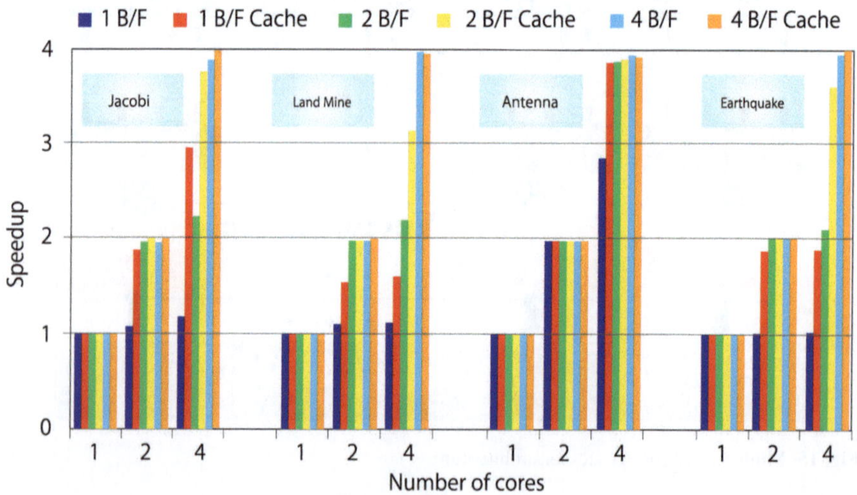

Fig. 16 Performance of multi-vector core processor

the bypass mechanism from memory directory to the register file and the cache work complementarily together to provide data to the register file. As a result, the equivalent performance of the 4 B/flop system is eventually realized even with 2 B/F. The speedup ratios of the 1 B/F system with the cache are limited compared to those of the 2 B/F system. This is because the shortage of the memory bandwidth of the 1 B/F system for the executions of these applications. Accordingly, these results suggest that the off-chip memory bandwidth per core should satisfy at least 2 B/flop to achieve high scalability in the execution of these kinds of applications.

For multi-threaded programs of various difference schemes, a thread can reuse the same data loaded previously by another thread. Let's look at the code shown in Fig. 13 again to discuss the locality of reference among cores. The outermost loop, index k, is parallelized, and arrays $H\_y(i, j, k-1)$ and $H\_x(i, j, k-1)$, which are not reused on a single thread, are reused as the data of $H\_y(i, j, k)$ and $H\_x(i, j, k)$ on another core in multi-core processing. Figure 17 shows the cache hit rates and the improved efficiency per core by the effect of the cache mutually preloaded among cores. Therefore, more cores are introduced, more higher hit rates are obtained. As a result, the performance gains per core increase as the cache hit rate increases. This is due to the prefetching effect among cores on the shared cache. These results indicate that many scientific and engineering applications designed based on the difference scheme are effectively processed on the CMVP with the vector cache.

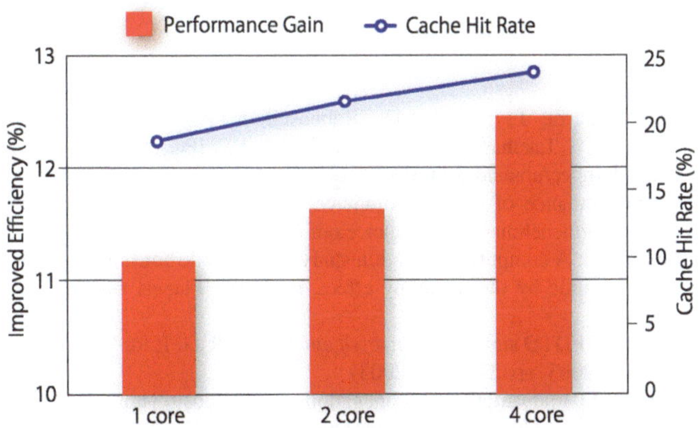

Fig. 17 Effect of vector cache for multi-vector core processor

6 Summary

This paper has discussed the performance of SX-9, the latest vector system at Tohoku University. The SX-9 system marked 19 top one scores out of 28 test of the HPC Challenge benchmark. The SX-9 system have shown the highly efficient execution in the Global FFT test, compared with pete-flop/s level scalar systems. We have also discussed code tuning case studies for SX-9, especially from the view point of effective use of ADB, newly installed on-chip software controllable cache. As data from ADB is available at the rate of 4 B/F, the key point for covering the limited off-chip memory bandwidth of SX-9 is how data with the higher locality is effectively specified in the code for its ADB caching. Through the experiments by using practical applications, we have confirmed the effectiveness of ADB to improve the sustained performance of SX-9. At the moment, users have to specify vector data with higher reuse possibility based on their experiences, however, smart tools to find out such portions are definitely desired. Finally, we have presented a multi vector-core design as the next generation vector computing. We have shown that vector caching is very effective to handle vector data with the intra-thread and inter-thread locality of reference, and is a key component of a multi vector-core processor to exploit its potential.

Acknowledgements This work has been done in collaboration between Tohoku University and NEC, and many colleagues contribute to this project. We would also like to thank Professors Akira Hasegawa, Kunio Sawaya, Motoyuki Sato, Satoru Yamamoto, Yasuhiro Sasao, Masahide Iizima of Tohoku University, Takahiro Tsukahara of Tokyo University of Science, and Dr. Keisuke Ariyoshi of JAMSTEC for providing the codes for the experiments.

References

1. Luszczek, P., Dongarra, J., Koester, D., Rabenseifner, R., Lucas, B., Kepner, J., McCalpin, J., Bailey, D., Takahashi, D.: Introduction to the HPC challenge benchmark suite. http://icl.cs.utk.edu/hpcc/ubs/index.html
2. Dongarra, J.: Performance of various computers using standard linear equations software. http://www.netlib.org/benchmark/performance.pdf (2009)
3. Katoh, Y., Iizima, M.: A computer simulation study on the mode conversion process from slow X-mode to fast X-mode by the tunneling effect. In: Earth, Planets and Space, pp. e53–e56 (2006)
4. Kobayashi, T., et al.: FDTD simulation on array antenna SAR-GPR for land mine detection. In: Proceedings of SSR2003, pp. 279–283 (2003)
5. Ariyoshi, K., et al.: Spatial variation in propagation speed of postseismic slip on the subducting plate boundary. In: Proceedings of 2nd Water Dynamics, vol. 35 (2004)
6. Takagi, Y., et al.: Study of high gain and broadband antipodal Fermi antenna with corrugation. In: Proceedings of 2004 International Symposium on Antennas and Propagation, vol. 1, pp. 69–72 (2007)

BSC-CNS Research and Supercomputing Resources

David Vicente, Javier Bartolome

Abstract Early in 2004 the Ministry of Education and Science (Spanish Government), Generalitat de Catalunya (local Catalan Government) and Technical University of Catalonia (UPC) took the initiative of creating a National Supercomputing Center in Barcelona. BSC-CNS (Barcelona Supercomputing Center—Centro Nacional de Supercomputación) is the National Supercomputing Facility in Spain and was officially constituted in April 2005. BSC-CNS manages MareNostrum, one of the most powerful supercomputers in Europe, located at the Torre Girona chapel. The mission of BSC-CNS is to investigate, develop and manage information technology in order to facilitate scientific progress. With this aim, special dedication has been taken to areas such as Computational Sciences, Life Sciences and Earth Sciences.

1 Overview

The key mission of the Operations Department is to ensure the continued availability and accessibility of RES systems 24 hours a day, 7 days a week and to provide support to all the users of the RES. Further core objectives are to manage upgrades to the MareNostrum and other RES nodes; facilitate access to RES facilities, including online electronic applications, remote access, and porting of code; manage the environmental aspect of the BSC-CNS installations; manage the technical aspects of integration of the MareNostrum in the DEISA and European HPC network grids; and ensure that RES staff receive appropriate training and skills development in order to be able to professionally carry out their duties in an environment of constant technological change and advancement.

David Vicente, Javier Bartolome
BSC-Centro Nacional de Supercomputacion, C/Jordi Girona 31, 08034 Barcelona, Spain, e-mail: david.vicente@bsc.es, javier.bartolome@bsc.es

M. Resch et al. (eds.), *High Performance Computing on Vector Systems 2009*,
DOI 10.1007/978-3-642-03913-3_2 © Springer-Verlag Berlin Heidelberg 2010

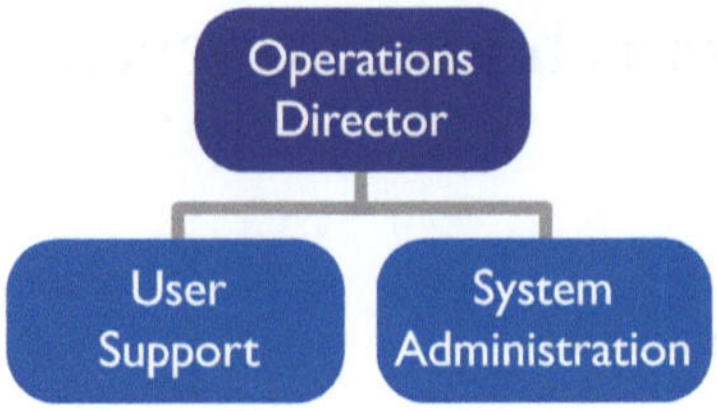

Fig. 1 Organisational structure

The Operations Department is structured in two divisions, System Management and User Support.

Systems Management supervises the daily operations of two key resources, the MareNostrum Supercomputer and the Spanish Supercomputing Network (RES), bearing responsibility for system administration, security, resource management, networking and helpdesk. This group also takes care of running all the other IT equipment installed at the BSC-CNS and related facilities.

User Support is responsible for direct user support providing detailed knowledge of programming models, libraries, tools and applications, and also is responsible for management of the BSC website, including the electronic management of access applications.

2 Supercomputing Resources at BSC

2.1 MareNostrum

In March 2004 the Spanish government and IBM signed an agreement to build one of the fastest computer in Europe. In July 2006 its capacity was increased due to be the large demand of scientific projects.

MareNostrum is a supercomputer based on processors PowerPC, the architecture BladeCenter, a Linux system and a Myrinet interconnection. These four technologies configure the base of an architecture and design that will have a big impact in the future of supercomputing. See below a summary of the system:

- Peak Performance of 94.21 Teraflops
- 10240 IBM Power PC 970 MP processors at 2.3 GHz (2560 JS21 blades)
- 20 TB of main memory
- 390 + 90 TB of disk storage
- Interconnection networks: Myrinet and Gigabit Ethernet
- Linux: SuSe Distribution

MareNostrum has 44 racks and takes up a space of 120 m^2, from these MareNostrum has 31 racks dedicated to performing calculations. These racks have a total of

10240 IBM 970 MP Power PC 2.3 GHz processors and 20 TB of total memory. Each computation rack is composed of 6 blade centers. Each rack has 336 processors and 672 GB of memory with an estimated peak performance of 3.1 TFlops. The JS21 nodes are grouped in a blade center which contains 14 dual processors with a total of 56 processors. Each blade center has two redundant power supplies. If one of the power supplies were to fail, the other power supply would keep the blade center running. Each blade center also has several switches for the interconnection with the several networks.

Each blade center has 14 JS21 server blades. Each of these nodes has two 2.3 GHz 970 MP PowerPC processors with 8 GB of shared memory between both processors and a local SAS disk of 36 GB, which is only used for swap and temporary data. Each blade boots by network and is able to work diskless with an IBM technology called DIM, that manages the OS images and their deployment. The 970 MP PowerPC processor has a 64-bit architecture designed for general use (the Mac G5 contains this type of processor). It is a superscalar processor with vector extensions SIMD (VMX) whose design is based on the high provision Power4 processor. The new 970 MP PowerPC processor is based on a 90 nm technology that works at a frequency of 2.3 GHz. It is ready to launch a maximum of 4 instructions per cycle and start up to 200 in-flight instructions.

Each node has a M3S-PCIXF-2-I Myrinet network card that connects it to the high speed network and the two connections to the Gigabit network.

The 2560 JS21 blade nodes are connected through a Myrinet high speed network. The different nodes are connected via fiber optic cables. Four of the 44 racks in MareNostrum are dedicated to networking which allow for the interconnection of the different nodes connected to the Myrinet network. These four racks are located in the center of the space and each node has its own fiber optic cable. The network elements connect the different cables allowing the interconnection from one point to another from the different nodes.

The total Myrinet interconnection network consists of the following:

- 10 Clos 256 + 256 switches
- 2 Spine 1280 switches

In addition to the 36 GB of capacity of the local disk of each node, MareNostrum has 20 storage servers arranged in 7 racks. These have a total of 560 SATA disks of 512 GB and each one provides a total capacity of 280 TB of external storage. These disks are working with GPFS (Global Parallel File System) which offers a global vision of the file system and also allows a parallel access. The 2560 nodes access the disks through the Gigabit interconnection network. Each of the 20 storage nodes has two p615 nodes to handle disk requests, a DS4700 controller and one EXP810 unit.

2.2 MareNostrum Performance 2008

MareNostrum has been in production the whole year, except for the scheduled maintenance periods. Using the remaining part of the year as the basis for calculations, the observed system utilisation is approximately 80%.

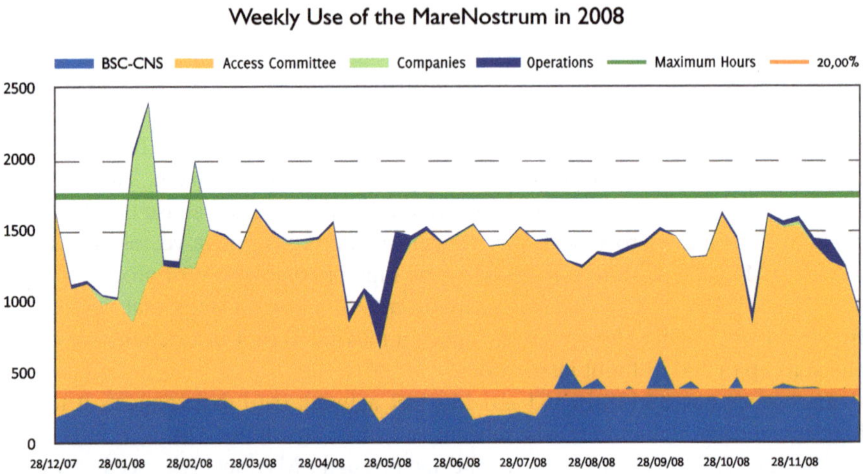

Fig. 2 Weekly use of MareNostrum in 2008 (in thousand of hours)

2.3 Shared Memory System

In 2007, BSC installed an SGI Altix 4700 shared memory system, allowing researchers to perform specific pre and post-processing analysis that require large shared memory systems. The system was upgraded twice in 2008, doubling its computational capacity, and increasing by 5 its memory capacity, enabling researchers to perform more complex analysis and work with larger data sets. The final configuration for this system is the following:

- 128 cpus Dual Core Montecito(IA-64)
  Each one of the 256 cores works at 1.6 GHz
- 8 MB L3 cache and 533 MHz Bus
- 2.5 TB RAM
- Peak Performance: 819.2 Gflops
- 2 internal SAS disks of 146 GB at 15000 RPMs
- 12 external SAS disks of 300 GB at 10000 RPMS

2.4 Backup and HSM Service

Due to the high demand for storage at BSC, coming from MareNostrum and other storage systems an SL8500 library was installed. During 2007 tape technology was migrated from LTO3 to LTO4 increasing the total storage capacity to 6 PB (6000 Terabytes) without compression. A total of 23 tape drives and 7500 tapes, each with 800 GB capacity, were installed to provide this storage. The SL8500 library in conjunction with 5 servers and 31.4 TB of disk space were configured and installed to perform two main services: backup of all compute systems at BSC and Hierarchical Storage Management (HSM).

A backup service using Tivoli technology was deployed permitting a daily backup of all data from MareNostrum and other systems. Three dedicated servers were installed and configured with the following backup roles:

- TSM1: MareNostrum Home directories
- TSM2: Tape management + MareNostrum project and applications directories
- TSM3: BSCServers + Laptops

Hierarchical Storage Management system, using SAMFS technology, permitted us to increase the storage capacity of MareNostrum up to Petabytes. HSM is a storage technology that provides automatic data migration from a costly device to one with lower cost. A storage of 9 TB of FiberChannel disks serves as costly or first level of storage and the SL8500 tape library serves as the second level of storage. Tools were provided to MareNostrum users for the movement of data between HSM and MareNostrum local filesystems.

2.5 Spanish Supercomputing Network

The Spanish Supercomputing Network (Red Española de Supercomputacíón RES) was inaugured in March, 2007 responding to the scientific community's need for increased calculation capacity. Currently RES provide more than 25 millions of cpu hours each 4 months.

In November 2006, BSC and IBM upgraded MareNostrum substituting the JS20 bladecenters for the JS21 and allowing us to double calculation capacity. Following this, the old MareNostrum hardware was used to create a distributed infrastructure of supercomputers in different sites in Spain, So RES consists of a distributed structure of supercomputers in order to give support to the supercomputing needs of the different research groups in Spain.

Access to RES resources:

The RES is accessible to users via a single access committee, with 44 scientist organized into four different panels. The calls for new projects are every four months and distribute more than 20 millions hours.

2.6 PRACE Prototype

BSC as a PRACE principal partner hosts one of the six prototypes for Petaflops systems. The machine called MariCel has been installed during the second half of 2008. Is a heterogeneous mix of Power 6 based JS22 nodes and Cell based QS22 nodes, both from IBM.

The summary specification of the prototype is:
Peak Performance of 15.6 Teraflops (double precision) from QS22 compute nodes.

- 72 QS22 nodes, each containing 2 PowerXCell 8i processors. 1296 cores
- 12 JS22 nodes, each containing 2 Power6 processors. 48 cores
- 960 GB of main memory
- 13 TB of disk storage

Interconnection networks:

- InfiniBand 4×DDR
- Gigabit Ethernet

The PowerXcell 8i processor is a 65 nm chip giving 103 Gflops performance and contains a mixture of 2 different elements specialised for their different roles. The PowerPC Processor Element (PPE) contains a 64-bit PowerPC Architecture core which can run 32 or 64 bit operating systems and applications. The 8 Synergistic Processor Element (SPE) are independent processing elements optimised for running compute intensive SIMD applications or threads and not operating system tasks. The SPEs depend on the PPE to run the operating system and the top level thread control and the PPE depends on the SPEs to provide the application performance.

The prototype design introduces the JS22 POWER6 nodes to allow system and cluster tasks to be offloaded from the QS22s to provide more resources for the computational tasks. The JS22s are characterized as a services nodes an arranged with 1 JS22 for 6 QS22s, occupying half of each Blade Center. This logical grouping is called a hypernode. The end user only sees a cluster of QS22 Cell based nodes to run jobs. The batch system software used is MAUI Cluster Scheduler running on top of SLURM as resource manager.

Each QS22 runs a Fedora 9 distribution with a kernel (v.2.6.25) customized by IBM and BSC in order to obtain the best performance from the Cell architecture over the IB network. A Red Hat Enterprise Linux 5.2 distribution is currently running on each JS22. All the nodes mount the operating system over NFS, using Distributed Image Management (DIM), the same technology that is being used in MareNostrum, to manage the different Operating System images. The Gigabit Ethernet network, configured with a Force10 S50N 48 port switch/router, is the one that runs the NFS, GPFS, job scheduling, and network booting traffic. The InfiniBand network, configured with a Voltaire ISR2004 96 port switch, directly connects the 84 compute and service nodes, and is used only for MPI communication. This network runs at 4 × DDR giving a maximum theoretical point to point bandwidth of 20 Gb/s (16 Gb/s of data).

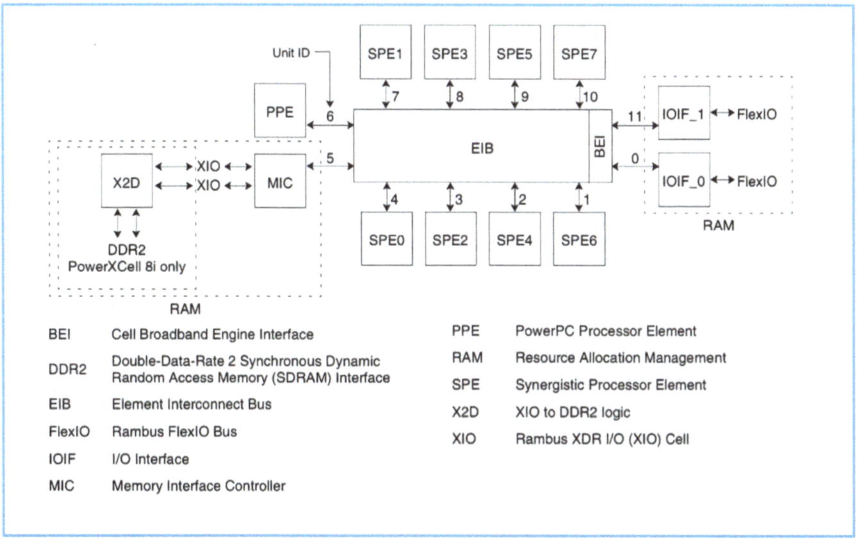

Fig. 3 PowerXCell 8i high level block diagram

Fig. 4 Maricel PRACE prototype

3 Research at BSC

The fourth main research areas at BSC-CNS are Computer Sciences, computer applications in Science and Engineering, Life Sciences and Earth Sciences:

Computer Science: The main objective of the department is to make advance in the hardware and software technologies available to build and efficiently use supercomputing infrastructures, bridging the gap between computer architecture and application requirements. The department is proposing novel architectures for processor, memory hierarchy and their interconnection, programming models and their efficient implementation, tools for performance analysis and prediction, resource management at different levels in the system (processor, memory, and storage) and for different execution environments, including Grid and e-Business platforms, as well as application optimization.

Life Science: The main goal of this department is to gain a deep insight into the origin, evolution and function of living organisms using theory and computation. Our focus goes from a global analysis of living entities understood as complex systems to detailed studies of key interactions at the sub-atomic level. Overall, our objective is to make theory and simulation as one of the driving forces to the advance of research in life sciences.

Earth Science: This Department was established with the objective of carrying out research in Earth system modelling, initially focusing on atmospheric physics and chemistry. The group has as main topics of research: air quality modelling, mineral dust modelling, meteorological modelling and global and regional climate modelling.

Computer Applications on Science and Engineering: The main goal is to enhance the links between two major contemporary fields of science and technology: applied science community and high performance computing experts. On one hand, the applied science community shows an increasing need of computing power. On the other hand, the high performance computational science experts have the knowledge and experience to get the best from supercomputing facilities. The evident beneficiaries of this interdisciplinary collaboration are: High-tech industry and Academia.

Challenges and Opportunities of Hybrid Computing Systems

Stefan Wesner, Sabine Roller, Harald Klimach

Abstract The advent of specialised processors and the movement towards multi-core processors lead to an increased complexity for the application developers as well as for computing centres aiming to deliver computing resources for a wider community. While the increasing variety of different computing architectures is typically seen as a challenge, this article shows—based on an application example of a coupled aero-acoustics simulation—the opportunities for energy and cost efficient simulations. Additionally the results are put in context of the prototyping activity of the Partnership for Advanced Computing in Europe (PRACE), and the potential benefits of a hybrid computing systems for compute service providers are discussed.

1 Introduction

A variety of different architectural approaches for supercomputers have been established over the years. Concrete installations are driven by the targeted user community and the research challenges to be addressed by the computing system. Additionally it is important whether the computing system is supposed to realise the best possible service for a specific community or whether a wide range of different communities need to be supported. It cannot be said that one system is better than the other as different communities demand very different configurations and architectures. The availability of multicore processors, heterogenous processors and accelerators of different types in increasingly short timescales have made the challenge to offer the "right" computing system even more complex.

On the other hand, the concept of integrating applications representing several aspects of a problem into one coupled simulations on one single system seems not to fit with the approach of a large homogeneous system, since the characteristics and

Stefan Wesner, Sabine Roller, Harald Klimach
Höchstleistungsrechenzentrum Stuttgart, Nobelstrasse 19, Stuttgart, Germany, e-mail: wesner@hlrs.de, roller@hlrs.de, klimach@hlrs.de

M. Resch et al. (eds.), *High Performance Computing on Vector Systems 2009*,
DOI 10.1007/978-3-642-03913-3_3 © Springer-Verlag Berlin Heidelberg 2010

with that the architectural requirements of the different parts of the simulation are distinct and do not usually fit to the same type of hardware. Such coupled simulation approaches are investigated in different application domains such as the Virtual Physiological Human (VPH) or in multi-scale and multi-physics applications like the aero-acoustics presented here in this article. The ideal environment for coupled simulations would be a system that tightly integrates different architectures into a single computing environment, a hybrid computing system, while keeping the latency low and the communication bandwidth high between the different hardware partitions.

Definition 1. A hybrid computing system is an environment that tightly integrates different architectures within a homogenous software environment from system software (operating system, scheduling system) over programming environments (e.g. MPI, OpenMP, ...) up to the tool (debugger, performance analyser, ...) level. It is extensible and allows different lifecycles for the different architectural parts realising a continuous upgrade path.

A general view on a hybrid computing system is shown in Fig. 1. The different hardware architectures (in this example scalar processors, vector processors, general purpose graphics processing units and accelerators) are integrated in one single system, all giving the same view to the operating system. Within such a system one could support not only different individual applications like e.g. Computational Fluid Dynamics (CFD), Computational Structure Mechanics (CSM) up to Molecular Dynamics (MD). These individual simulations have been "the application" in the past. But nowadays, they become modules in more complex applications. Several individual codes representing different physics are combined to come closer to reality. On the one hand side, there is a need for multi-physics applications since nature typically is not single-, but multi-physics, while on the other hand it is possible now to run multi-physics applications since algorithms and codes are mature enough and computers fast enough.

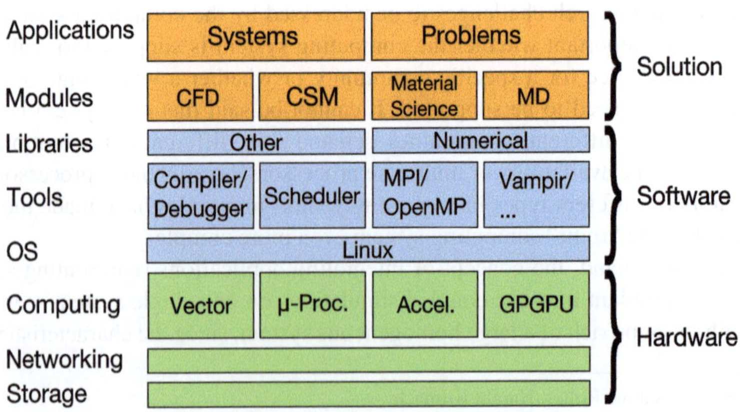

Fig. 1 The conceptual view on a hybrid computing system

As of today such a system must be seen as hypothetical as there are no vendors offering such a configuration fully. The different architectures are typically not only different in processor architectures, but also in operating systems (even if Linux is widely available, many different flavours are in use), network interconnect and the whole software stack on top of the operating system. Additionally most scheduling systems do not support scheduling across "architectural" boundaries.

2 European Context

Within the European infrastructure research project Partnership for Advanced Computing in Europe (PRACE) several prototypes are investigated in order to validate potential future multi petaflops computing system solutions. The motivation for the prototyping activity is the large financial risks associated with investments in multi petaflops systems and the need to minimize technological risks by validation of candidate configurations. PRACE brings together the large computing centres in Europe aiming for an alignment of activities on technical level but similarly on the organisational level with the goal of realising a world wide competitive high end computing infrastructure in Europe. The goal is to establish three to five European Supercomputing centres (also known as tier-0 centres) realising a performance pyramid supporting users from regional centres (tier-2) over national computing centres (tier-1) up to the leading edge computing systems.

In Germany, the GAUSS Centre for supercomputing is participating in PRACE and is one of the candidates for hosting a tier-0 computing system. The GAUSS centre brings together the Jülich Supercomputing Centre (FZJ), Leibniz-Rechenzentrum der Bayerischen Akademie der Wissenschaften in Garching (LRZ) and the Höchstleistungsrechenzentrum Stuttgart (HLRS). On the regional level the GAUSS Alliance is providing the tier-2 infrastructure for Germany. Similar approaches are followed in other European countries.

Within PRACE several prototypes had been selected as shown in Table 1 with a large variety of different architectures. The HLRS/NEC prototype is a first instantiation of the hybrid computing system concept as outlined in the previous section. The HLRS/NEC prototype is realised using three different architectures (Vector proces-

Table 1 PRACE prototypes for petaflops systems 2010/2011

Host	Country	Architecture	Processor
FZJ	Germany	MPP	IBM BlueGene/P
CSC-CSCS	Finland & Switzerland	MPP	Cray XT5/XTn - AMD Opteron
CEA-FZJ	France & Germany	SMP-TN	Bull et al. Intel Xeon Nehalem
NCF/SARA	Netherlands	SMP-FN	IBM Power 6
BSC	Spain	Hybrid (fine grain)	IBM Cell
HLRS	Germany	Hybrid (coarse grain)	NEC Vector SX-9 + Intel Xeon Nehalem + NVidia Tesla-10 GPGPUS

sor, Scalar Processor and GPUs) and needs to interconnect the different parts using different network technology, the NEC-IXS interconnect and Infiniband.

Table 2 HLRS/NEC prototype configuration details

Architecture	#processors	#cores	Peak Performance
NEC SX-9 Vector	192	n/a	19.2 TF
Intel Xeon Nehalem	700	5600	62.7 TF
NVidia Tesla-10	16 units/64 cards	15360	64 TF

3 Validation Scenario

In order to validate the hybrid computing systems approach a coupled aero-acoustic simulation has been chosen as application scenario. Direct simulation of aero-acoustics is a typical multi-scale problem, as the noise is generated in small flow structures, but propagated over a long distance. Usually the noise generating flow happens around structures with arbitrary complex geometry. Thus, there is the need for unstructured meshes around these structures. Sound wave propagation on the other hand mostly spreads over a large domain, which can be simply discretized with a structured Cartesian grid. As unstructured meshes are more costly than structured meshes, the aim of the domain setup is, to switch to structured meshes as soon as possible. In Fig. 2 a typical domain decomposition is shown. There is a unstructured domain around a gas injector nozzle with fine grained resolution, where the actual flow structure is resolved. Shortly after the outlet of the nozzle the discretization is switched to a fine grained structured grid, which is neighboured by coarser structured grids, in regions where no high resolution is needed anymore.

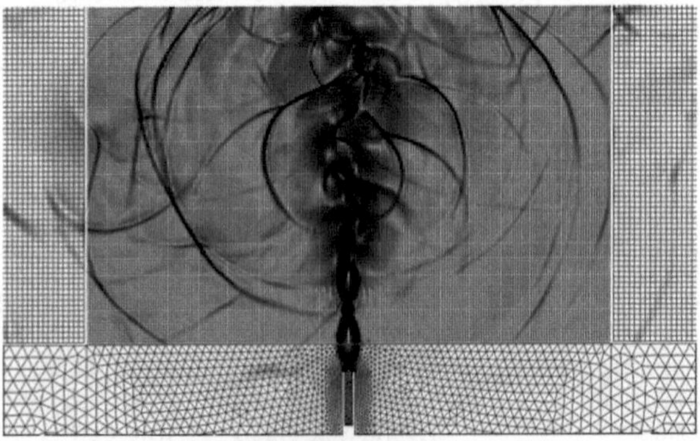

Fig. 2 A simplified version of the hybrid systems validation scenario

The deployed coupling scheme allows different time steps in adjacent domains by a subcycling mechanism. This is necessary, as different spatial resolutions imply also different temporal resolutions. Furthermore the coupling scheme is capable to handle different underlying equations to model the physics in each domain. Due to all these physical properties, each of those domains has its own numerical requirements.

The goal of the scenario is to show that if the problem is distributed in a way that the structured and unstructured parts are executed on their best fitting architecture, an advantage in speed and efficiency can be gained.

The goal is the full 3D simulation of a nozzle with four outlets of 0.15 mm diameter with an unstructured domain of $8 * 8 * 10\ \text{mm}^3 = 15$ million elements embedded into a structured domain of $2 * 2 * 5\ \text{cm}^3$ with 1 billion cells and a coarsely resolved far field (also structured) of $1 * 1 * 1\ \text{m}^3$ with 350000 cells (Fig. 3. The initial guess for such a simulation using a hybrid setting of 1024-2048 Intel Xeon Harpertown plus 128 CPUs NEC SX-9 gives a need of elapsed computing time of 5 days. The same simulation on one single architecture (either micro-processor or vector) would be just impossible, as will be seen later on.

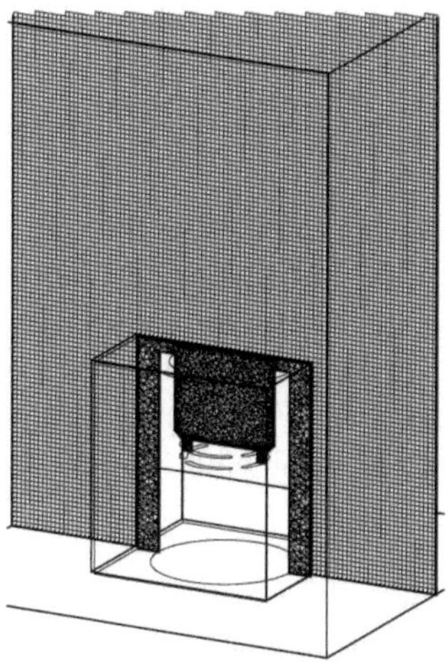

Fig. 3 A more sophisticated hybrid systems validation scenario

4 Initial Results

Initial simulations on the hybrid prototype have been carried out with a setting equivalent to the description above, but the structure domain reduced to a very tiny domain. This has been necessary to set up and configure the hybrid simulation. Here, the structured domain which vectorizes very well and thus runs most efficient on the vector part of the hybrid prototype, the unstructured domain is very expensive on the vector system due to its massive usage of indirect addressing and pointer structures and is therefore run on the scalar part of the system.

From the values in Table 3 it can be clearly seen that the two parts of the simulation perform very differently on the two involved types of machines. Especially the misfit of the unstructured domain on the vector system is obvious, but also the much better performance of the structured domain on the vector system. However, due to the very small structured domain, the benefits gained on the vector system is not high enough to make the hybrid approach more attractive than the X86-only run. The testcase resp. its most important part is simply to small to make the usage of an HPC system efficient.

Table 3 Initial simulation results with very small testcase

Program part	Intel Xeon Nehalem 37 structured 1 unstructured	SX-9 1 structured 1 unstructured	Coupled structured: 1 SX-9 unstruct: 1 Nehalem
UNSTRUCT	101.47	**1128.36**	101.47
STRUCT	**317.79**	30.33	30.64
KOP	2881.68	106.43	117.86
waiting	93.42	992	
Total elapsed (s)	98.07	1234.79	134.66
Price in EUR	0.21	4.46	0.23

Nevertheless, 2 different topics in choosing the appropriate distributions can already be seen. The one is the goal of efficient load balancing. On the X86-only system, the structured part is already using a much The memory consumption is a main issue for this problem. On the Intel Nehalem already 5 nodes have to be used to host the complete structured domain. For larger structured domains this increasingly becomes a problem, and is an advantage for the NEC SX9 with its large available main memory per process.

In the next step, the structure domain is increased by a factor of 10, which is still small While being still a comparably small test case the impact of distribution the work to the corresponding best suited architecture can be already seen. As shown in Table 4 the pure vector case is still dominated by the execution time for the unstructured part but shows even a slight decrease in execution time and price. The pure scalar case suffers from the ten times increased execution time of the structured part and shows a massively increased overall execution time and corresponding increase in price. The hybrid case shows a much better robustness against the change in the configuration and benefits from the short execution time of the unstructured domain on the Nehalem cluster and the fast execution of the structured part on the SX-9 system. The increase in price by a factor of roughly 3.5 for a ten times increased structured area demonstrate impressively the potential of the approach for the chosen test case.

Table 4 Estimated results for a ten times increased structured domain

Program part	Intel Xeon Nehalem 37 structured 1 unstructured	SX-9 1 structured 1 unstructured	Coupled structured: 1 SX-9 unstruct: 1 Nehalem
UNSTRUCT	101.47	**1128.36**	101.47
STRUCT	**3177.9**	303.4	303.4
KOP	2881.68	106.43	117.86
waiting	92.42	700	
Total elapsed (s)	3177.92	1128.36	395.85
Total elapsed (h)	0:56'	0:26'	0:10'
Price in EUR	6.71	4.07	0.72

5 Operational Requirements

The provision of a hybrid computing system imposes also challenges on the provider side. In order to allow the benefits of using diverse architectures realise in coupled simulations several conditions have to be met including:

1. Different architectural parts need to be connected with high bandwidth and low latency
2. Jobs across system parts should be scheduled as a virtually single job
3. Consumed resources on the different parts are tagged as related allowing the accounting (and potentially further on billing) of two or more simulation parts together
4. Performance analysis, debugging and other developer tools need to be able to operate across different hybrid system parts

In an ideal case the different parts would be connected using one single technology. However this would limit the potential parts of the hybrid systems based on the available interconnect technology and would constitute a too restrictive constraint. Consequently a typical hybrid system will have to allow the most effective interconnect technology within a system part and will seek for the lowest denominator (such as 10 Gigabit Ethernet) to connect the different parts or would aim to directly integrate the different parts within the nodes such as using the PCIExpress interface. So in general the requirement of high bandwidth and low latency between the different parts cannot be met and one need to accept even a variety of connectivity parameters across the overall system. For the HLRS/NEC prototype within PRACE a combined solutions relying on NEC-IXS, Infiniband and Gigabit Ethernet has been realised as shown in Fig. 4.

Scheduling jobs across system parts allowing a co-scheduling of jobs starting at the same time on the different parts would require a meta-scheduling or integrated scheduling system allowing a user to specify resource demands for the different parts (two or more) and submit one single job description covering the different application parts and all steps to start and wrap-up the simulation. Additionally it has to be considered that on a hybrid system not all jobs will be hybrid and need to co-exist with regular single architecture jobs. Combining this with requirements on

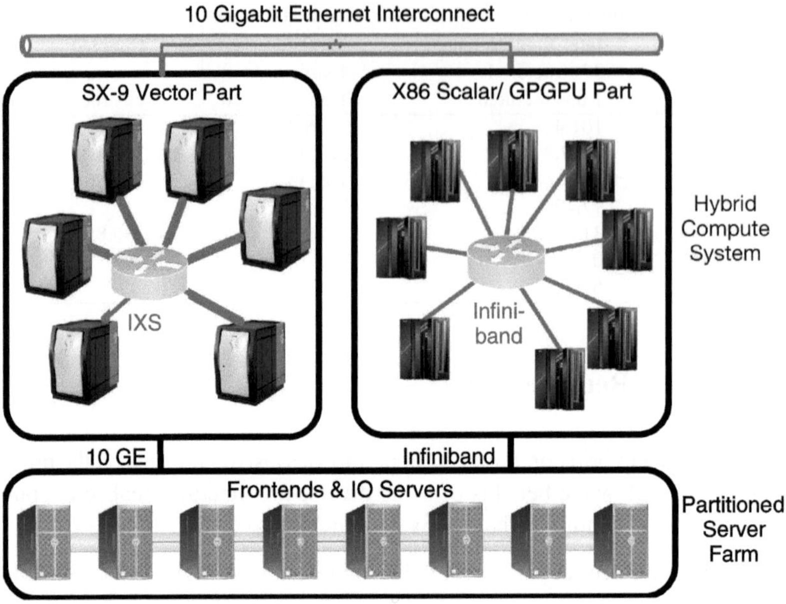

Fig. 4 HLRS/NEC prototype physical architecture

future scheduling systems to support also different class of users and corresponding Service Level Agreements (SLA) it also has to be stated that these requirement cannot be met as of today.

The third aspect of cross system parts accounting e.g. by tagging different job parts as related is mostly prevented from the fact that not even a single operating system can be assumed across the different system parts. Different operating system uses quite different methods for collecting resource utilisation data and most of them do not even support a tagging of several operations within one job with an accounting id or a similar mechanism.

Using developer tools across system boundaries would be needed to support the application developer during the software development process and similarly to identify performance bottlenecks in particular for finding the right balance between the different system parts. The availability of tools supporting hybrid application development are not available as of today.

While the development of applications for hybrid computing systems is still in its infancy the situation for the system operation is even worse. In order to target a wider take up of hybrid application programming the user environment need substantial improvement starting from cross system part scheduling up to developer tools support. From the computing centre provider viewpoint the potential need to acquire a hybrid system from different vendors relying on different operating systems and network technologies and an increased complexity for accounting and billing of inhomogenous system require an increased effort for integration and operation.

6 Conclusions and Future Work

The initial results presented here and previously in REF ECCOMAS show clearly the potential of hybrid computing systems in order to reduce the overall execution time of a simulation and similarly to reduce the costs and carbon footprint of a simulation. However the complexity of realising such a coupled simulation need to be balanced against the speed and/or cost benefit that can be achieved. In particular for comparably small problems and where the different coupled parts show a similar affinity to a certain compute system architecture the effort is not justified.

However for very large problems where the application part operating on the *wrong* architecture shows an increasing negative impact on the overall application performance and limit the problem size that can be tackled the effort for realising the hybrid application is justified. Consequently future research should not only focus on realising specialised coupled simulations (e.g. for realising leading edge performance) but aim to support the development of applications exploiting different architectures using standardised libraries and primitives where certain compute intensive *kernels* can be kind of outsourced to other parts similar to approaches chosen for CUDA [2] or OpenCL [1].

The user environment that can be delivered on prototypical hybrid computing systems is not appropriate for a daily use as of today. Additional work is needed to ease the application development and execution of hybrid application similarly to single system jobs. Additionally tool support is needed supporting the user to predict based on the specific problem case an appropriate distribution on the right amount of resources of the different types.

Acknowledgements The work presented here is partially funded by European Commision under contract number RI-211528. Contains partially pictures produced by Michael Resch (HLRS) and Claus-Dieter Munz, Jens Utzmann Institute for Aerodynamics and Gasdynamics (IAG)

References

1. Munshi, A.: The OpenCL specification. Technical report, Khronos OpenCL Working Group, (2009)
2. Nvidia, Cuda reference manual. Technical report, Nvidia (July 2009)
3. Roller, S., Klimach, H., Munz, C.-D., Utzmann, J.: Heterogeneous coupling in computational aeroacoustics. In: 5th European Congress on Computational Methods in Applied Sciences and Engineering (2008)

Going Forward with GPU Computing

Guillaume Colin de Verdière

Abstract This article describes why CEA is looking at GPU Computing and how the first experiments are conducted. We describe here a well defined global strategy which relies on training users and taking advantage of Grand Challenges, involving early access users and system administrators. We also describe some preliminary results and raise questions which need to be addressed in the near future.

1 Computing needs at CEA

The CEA (Commissariat à l'Énergie Atomique—the French Atomic Energy Commission) is in charge of R&D in fields such as energy (fission, fusion and other kinds of sustainable energy), life science, information technologies and defense. All these domains use numerical simulation and need more and more computing power. In some cases, the researchers are studying fields where it is impossible to carry on experiments (when, for example, trying to understand how the universe has evolved since the Big Bang or to describe in detail the behavior of matter in the most extreme conditions—see Fig. 1). All in all, the computing needs at CEA are almost limitless.

To keep up with its users' needs, CEA runs two large computing facilities, one for its R&D needs (named CCRT for Computing Center for Research and Technology) and one for the Defense programs (named TERA). These centers offer the best possible machines at a given date and try to evolve along with computer technologies. Therefore, it is CEA's role to study the potential benefits of new techniques and, based on the latter, design and buy new more powerful computers on a regular basis.

Table 1 shows the evolution of the CCRT and TERA machines in terms of total compute power. One should note that the increase of proposed power is meant to

Guillaume Colin de Verdière
CEA, DAM, DIF, 91297, Arpajon, France, e-mail: guillaume.colin-de-verdiere@cea.fr

M. Resch et al. (eds.), *High Performance Computing on Vector Systems 2009*,
DOI 10.1007/978-3-642-03913-3_4 © Springer-Verlag Berlin Heidelberg 2010

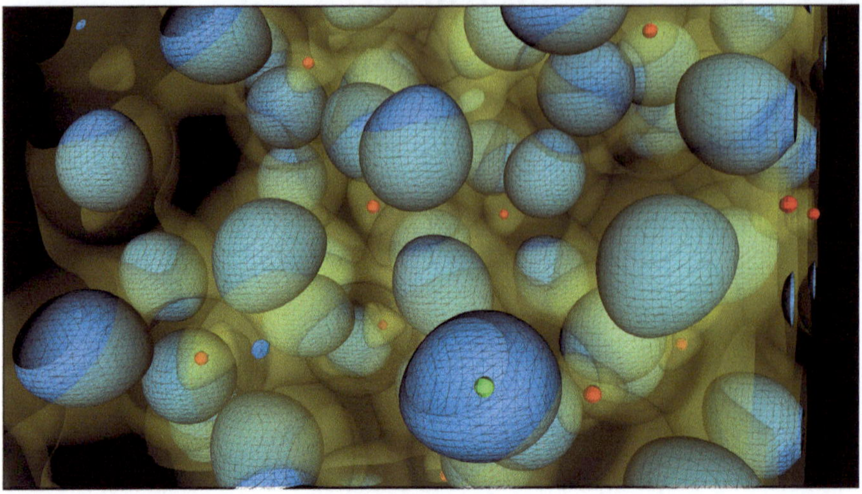

Fig. 1 Example of material simulation: behavior of SiO_2 under very high compression. This situation arises in high energy laser experiments such as those conducted in the LIL facility located at CEA Cesta. ©CEA/DAM

beat Moore's law: mere processor technology evolution will not answer our expectations.

Table 1 Compute power provided to users at CEA as of May 2009. In 2009, alongside the 52 TFlops Itanium machine of the CCRT, was added a 103 TFlops Nehalem based machine which includes a 192 TFlops (Single Precision) GPU extension which is described in the next table

	1996	2001	2003	2005	2007	2009	2010
CCRT			2 TFlops		52 TFlops	52 + 103 + 192 TFlops	
TERA	43 GFlops	6 TFlops		61 TFlops			over 1 PFlop

Following the trend described in Table 1, CEA should offer multi-petaflop systems in the near future. The next TERA machine is expected to be in production at the end of the year 2010 and should exceed a petaflop (peak). Users are already requesting even bigger machines. Thus the design of an exaflop system, or at least the studies on related challenges, has already begun since it takes many years to design such a machine in collaboration with the vendors and build or prepare the machine room.

Two main issues already arise : the electrical power required to run the future machine as well as the capabilities of current and future processors. Extrapolating the current trends in processor technology, we foresee that we will need more than 100 MW for the machine itself, without taking into account the power required for the infrastructure (UPS, cooling...). Note also that memory and interconnect will possibly dominate the power consumption in the near future. This is not an ecofriendly situation if one needs to build a power plant against one and every computer center around the world. To curb this trend, one hope would be a dramatic increase of individual processor capabilities with a minimal thermal dissipation. Unfortunately, no disruptive technology for such processors is yet foreseen. The frequency increase has almost come to a stall, the number of cores is increasing only slowly. Some even claim that the performance will in fact decrease if there are more core per processor than currently available.[1]

Some pieces of hardware show impressive performances while not being too power hungry. They are commonly implemented as attached processors (Cell, FPGA, Clearspeed boards, GPU...). Only the GPU is "mainstream" which is why it has attracted our attention.

The **G**raphic **P**rocessing **U**nits have been increasingly used for general purpose computing instead of just producing images, coining the term of GPGPU for General Purpose computing on GPU. Within a couple of years, they have featured increased performances while being easier to program and will progressively include full floating point compute units features (ECC, double precision).

GPU computing is therefore CEA's current answer for dramatically increasing performances while keeping the energy envelop under control.

2 Starting the Process

The introduction of a new technology must be planned ahead enough to get the users ready in time. A choice of sample machines should be available to them and training sessions set up. At the same time, implications of the new technologies must be studied and potential problems solved before this new technology can be turned into production mode (here we mean available to the broad community of users).

The following sections will cover each aspect of this initial approach.

2.1 Available Hardware

It is believed at CEA, that users should have test machines easily accessible. Therefore, we have provided small tests machines to learn how to program GPUs. A large visualization cluster was installed to see how post-processing tools could benefit from GPU computing, as well as start experimenting larger simulations using GPU computing. In the meantime, a large cluster, called Titane, has been installed to start testing hybrid programming (MPI + CUDA for example) and get prepared for running a large hybrid machine for production.

In other words, this set of platforms is for complementary purposes:

1. small test machines for GPU programming basics,
2. a visualization cluster for post-processing and moderate parallel hybrid testing,

[1] http://www.sandia.gov/news/resources/releases/2009/multicore.html.

3. a large cluster for full scale GPGPU computing in production conditions (shown in Fig. 2).

Fig. 2 Photo of the new 192 TFlops GPGPU machine of the CCRT. ©CEA/DAM

CEA chose to use GPUs from NVIDIA for two reasons: the Tesla is a 1U blade (see Fig. 3) which makes it easy to integrate in server racks. The software provided by NVIDIA was the most mature at the time of the decision of purchase. Figure 4 explains the logical design of one C1060 and Fig. 5 shows the actual T10 chip which is the heart of the C1060. Seeing the large number of units available explains why NVIDIA cards are able to handle many of concurrent threads. This number has consequences on the software side which will be discussed in Sect. 2.2.

Fig. 3 Inside view of a NVIDIA Tesla S1070 featuring 4 Tesla C1060 boards. ©NVIDIA Corporation

One of the test machines has even been labeled as a WP8 prototype of the PRACE project,[2] further validating the vision already implemented at CEA. This machine

[2] http://www.prace.eu.

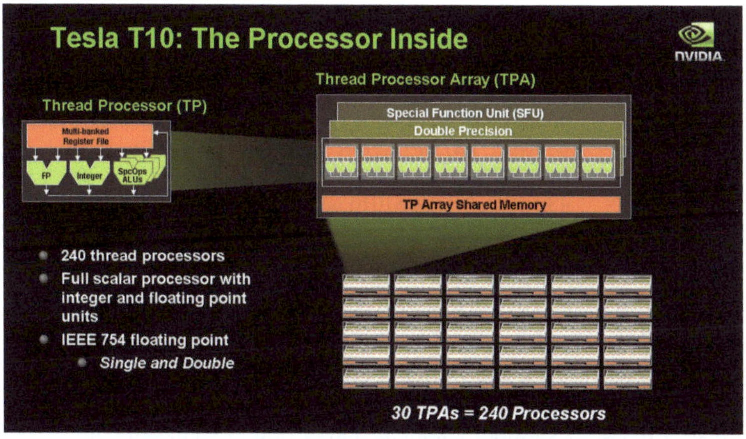

Fig. 4 Logical design of a Tesla C1060 multiprocessor. ©NVIDIA Corporation

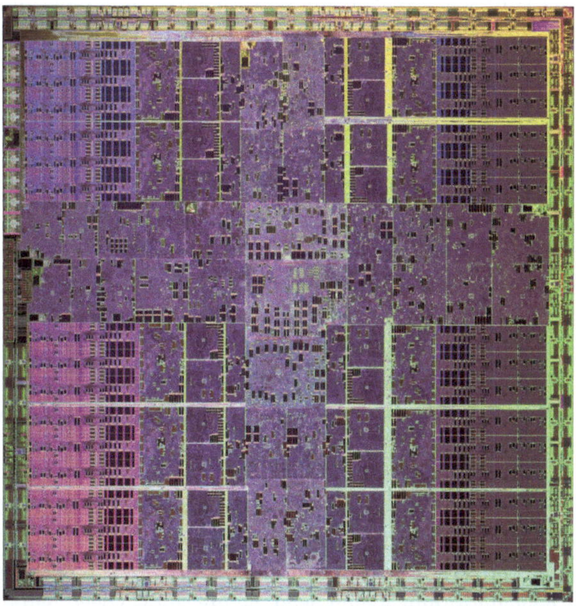

Fig. 5 View of a NVIDIA Tesla C1060 chip. Actual layout of the multiprocessor described in Fig. 4. ©NVIDIA Corporation

will be opened to the PRACE community and will serve as a testbed for programming and debugging environments.

Table 2 Different GPU enabled machines available at CEA/DIF for GPU computing

Machine	Description
Tests machines	2 BULL R425 servers + 2 NVIDIA Tesla S1070
Viz Cluster	40 HP nodes: NVIDIA Quadro FX 5800 + 8 cores *Haperton* + 64 GB
	96 BULL R422-e2 servers
Titane (GPU part)	Intel Nehalem-EP, 8 GB, IB DDR
	48 NVIDIA Tesla S1070 = 192 TFlops SP

2.2 Choosing a Programming Language

A hybrid machine is not a ready-to-use system in the sense that users have to adapt their codes to take advantage of such machines. The question which arises immediately is that of the programming language which should be promoted. Answers are obviously difficult to give in general and without experiments. The task assigned for the year 2009 is to select reasonable options and assess them in order to learn the pros and cons of each language or programming model. From the knowledge and lessons gained, advice, if not help, will be given to users to jump on the GPU computing bandwagon.

2.2.1 CUDA

The natural choice, when using NVIDIA GPUs, is to program using the vendor's solution. This is the way to get the most out of the graphic card. One downside is that the code has to be fairly or even heavily changed since CUDA is not 100% compatible with C or C++. On the up side, CUDA looks very much like C and is really straightforward to learn syntactically. Any newcomer to GPU Computing should try it anyway to understand the memory layout proposed by NVIDIA. This very model has been adopted by OpenCL and mastering it will also help to have efficient HMPP codes too. HMPP is described in more details in Sect. 2.2.4. All in all, the real issue is the memory model and management as well as SIMD thinking (in terms of thousands if not millions of threads).

From our first experiments, we would recommend using CUDA mostly for prototyping codes or new algorithms which need the uttermost speed while not requiring long term portability. It is also well suited for highly tuned libraries. In this case, a highly knowledgeable programmer will provide and maintain the library.

At the time of writing this article, the status of CUDA/FORTRAN is really unclear. The problem is that a great deal of physics codes used at CEA are still FORTRAN based. Coupling those with CUDA requires an extra workload which might be a big annoyance to users. And yet, the advanced users are willing to try CUDA. Should CUDA/FORTRAN be available, a large community will likely be attracted to GPU computing.

2.2.2 OpenCL

OpenCL is depicted as *the* solution to portable GPU computing. Supported by the Khronos Group, OpenCL has been designed from the CUDA model. Thus the memory model is the same. Many vendors have joined the design effort, but no implementation has been provided to this day. The actual portability of OpenCL codes has yet to be demonstrated, especially in terms of performances. We hope to be able to test it soon on AMD/ATI graphic boards, in order to introduce diversity in our approaches. We are also very interested in what kind of OpenCL support will be available on Intel forthcoming Larrabee.

One strong limitation of OpenCL is the lack of FORTRAN support.

2.2.3 RapidMind

Another well known player in the field of GPGPU is RapidMind. This technology will have to be experimented soon since it has proved useful in the finance world. A strong drawback of the RapidMind solution is the lack of FORTRAN solutions as well. Moreover, it is viewed as a very intrusive solution (specific API) but this has to be confirmed by further in depth tests.

2.2.4 HMPP

The HMPP workbench provided by *CAPS Entreprise*[3] is a smart alternative solution to program GPUs. It is our key solution for legacy codes since HMPP is compatible with C and FORTRAN. The ease of use is another feature. With such a tool, CEA will capitalize more easily on its millions of code lines. Being a directive based tool (#pragma or FORTRAN comment) on the same token as OpenMP, going out of HMPP is costless: a code will compile and execute as expected even there is no HMPP compiler.

The main advantages of the HMPP workbench for CEA are:

- FORTRAN and C support (see Figs. 6 and 7).
- Java, C++ support will be available soon.
- Multiple targets (NVIDIA, ATI, SSE, ...) are already offered by HMPP. This means that the programmer will only have to concentrate on parallelism at a higher level and not focus on idiosyncrasies of the underlying hardware.
- As already mentioned, HMPP truly keeps the codes' portability.
- HMPP has a low learning curve as described below.

Programming with HMPP does not mean that special features of the hardware have to be ignored completely. It will be necessary to have a good understanding of its limitations to get the best performances out of HMPP. For example: knowing that

[3] http://www.caps-entreprise.com.

the PCI-Express bus is a real bottleneck, it is worthless to offload computations to a GPU for a small amount of data. This is where conditionals will play an important role to fine tune the application at run time, as illustrated in Fig. 6.

```
#pragma hmpp lb1 codelet, args[v1].io = out, target = CUDA,
#pragma hmpp lb2 codelet, args[v1].io = out, target = SSE
#pragma hmpp lb3 codelet, args[v1].io = out &
#pragma hmpp lb3 codelet, cond = "n > 800", target = CUDA
void MyCodelet(int n, float v1[n], float v2[n], float v3[n])
{
  int i;
  for (i = 0 ; i < n ; i++) {
    v1[i] = v2[i] + v3[i];
  }
}
int main(int argc, char **argv)
{
  int n;
  ........
#pragma hmpp lb1 callsite
  MyCodelet (n, m, myinc, inm, myoutv); // CUDA version
  ........
#pragma hmpp lb2 callsite
  MyCodelet (n, m, myinc, inm, myoutv); // SSE version
  ........
#pragma hmpp lb3 callsite
  MyCodelet (n, m, myinc, inm, myoutv); // CUDA if n > 800
  ........
}
```

Fig. 6 Piece of code illustrating HMPP's usage of directives.[a]

[a] Depending on the label used at the callsite level, one of the different "codelet" will be used. A codelet is, in first approximation, the equivalent of a CUDA kernel. It is the portion of the code which will be offloaded to the GPU. The usage of an accelerated version could also depend on runtime decisions such as the one declared in the third codelet declaration.

```
PROGRAM simplepgm
!$hmpp simple callsite
  CALL simple(n,t1,t2,t3,alpha)
!$hmpp simple release   PRINT *, t1(1), t1(2) , " (...) " , t1(n-1), t1(n)
END PROGRAM simplepgm
!$hmpp simple codelet, target = CUDA
SUBROUTINE simple(n,v1,v2,v3,alpha)
  IMPLICIT NONE
! ....
```

Fig. 7 Snippet of code illustrating usage of HMPP in FORTRAN

2.2.5 A Remark on Languages

At the current stage of GPU computing dissemination, the question of the programming language is not that essential. The real challenge is to understand the new memory layout provided by this new class of hardware and how to use thousands of threads per compute unit. The next step is to quantify the actual impact on the code architecture. All this will take time. By the time programmers will be comfortable with this new way of thinking, the language landscape should have cleared some. They will then be able to start programming portable and long lasting programs.

2.2.6 Training Sessions

Once we had selected a set of languages (namely CUDA and HMPP), it appeared essential to us to create, as a first action, a community of early adopters who will master these new technologies. Since it is an entirely new way of thinking, it is required to ease the change of paradigm. As we noticed that, even within some of the most skillful programmers, parallelism is *not* a widespread knowledge, we have to prepare them to move to massive parallelism. Once they do, they will spread the knowledge within the teams. These early adopters are envisioned as catalytic seeds who will be able to talk with their fellow co-workers in their every day words. This process will take time (months if not years) but this is the only way to successfully introduce a disruptive technology in their habits.

We introduced GPU computing through tutorial sessions. We noticed that within 3 days for CUDA and 2 days for HMPP, users were up to speed and willing to experiment on their codes. The ease of use of HMPP as well as its potential has been clearly demonstrated during those training sessions.

To be sure that each participant got enough training, we limited the size of the groups to 12 students at most. To get their interest constant all along the training period, we mixed theory and hands-on sessions. The latter were based on real users' problems: for example, we had them build a convolution step by step.

The feedback we received from all these sessions was excellent. This encourages us to set up more of such training sessions in the future.

2.3 The System Administration Side

Introducing GPUs as a computing device in a production computing center (i.e. computing 24 hours a day, all year round), such as the CCRT, requires all implications to be anticipated and well understood by the system administrators (in short sysadmins). Therefore, we have a strategy based on Grand Challenges to help the sysadmins. Grand Challenges are computations requiring exceptional resources (generally the whole machine) to solve a new or difficult problem. They run on special time slots on a production machine. The following section describes how those runs are used from a special standpoint and will tackle the issue of predictable problems.

2.3.1 The Grand Challenges Strategy

The first few weeks of life of new machines are very important (especially for the largest ones). They will show the weakest parts needing to be replaced. During this period of time, machines can be a little unstable. Sysadmins get to know their actual features, the way to fine tune them, how to upgrade their software....

Two situations may arise: on one hand, a shy user may want to avoid this period of instability. On the other hand, some are ready to take the chance of having the full machine run exceptional computations, even if some fail due to the youth of the machine. To promote this behavior, we issue call for Grand Challenges to motivate users help us stress the machines. This is also a great incentive for them to port their favorite application to the new system.

Through Grand Challenges, the sysadmins can monitor the new machine in real situations of code development and of long and stressful runs. This is a far better way to get closer to real production conditions than just running a Linpack or a suite of small benchmarks. Sysadmins also have the opportunity to measure the actual power consumption with a representative workload.

2.3.2 Foreseen Problems

For our system administrators, it is capital to provide a system which is as stable as possible. On a very stable machine, users can launch runs lasting days if not weeks, as it is currently observed on CCRT machines. The goal is to have the same level of quality of service with a hybrid system.

Using a technology, which has mainly been developed for "gamers", in an HPC context raises some questions. The following points are the start of a list of items which should be addressed soon:

- Driver versions: for a gamer, frequent changes of driver version is a way to enjoy the latest game being released. In an HPC context, we cannot afford to change the machines' configurations constantly. Therefore, we will have to find out a process for qualifying releases of drivers while not lagging behind too many releases. It will be a balance between new features and stability.
- Hardware availability: how do we guarantee that, when a job starts, the hardware is present, in good shape and ready to accept workloads from the user? Adding this extra piece of hardware increases the complexity of the machine. Sanity checks will become harder in the future.
- Lack of ECC: to add more burden on the sanity check item list, there is one important aspect of current generation of GPUs: their lack of ECC. How can we guarantee a user that the machine is computing correctly and that not a single bit goes astray? Should they introduce check-sums in their code, thus introducing

unwanted complexity in them? Should we have small benchmarks that will be run before each user's job? What kind of notification will be provided to both the sysadmins and the users if a graphic board memory fails? We obviously have to answer these questions in details before being able to fully open the machine as a production class one while waiting for ECC to become a standard in future versions of GPU hardware.

- Impacts of time sharing: gamers are inherently using their machines alone. Only one instance of a program is accessing the hardware resources. Therefore, sharing GPUs between users can be tricky. What will happen to user 2 if user 1 exhausts most of the GPU memory? This will almost certainly lead to a crash of user 2's application. To avoid such race conditions, we will favor a single user per node for a starter.
- Security issues: CEA's facility is providing cycles to industrial partners. It is crucial to guarantee that *nobody* can spy on their computations. Every newcomer to CUDA has made the following experiment: a student runs a correct program on a GPU; a second one runs a program that (wrongly) does nothing but retrieves the results. To the second student, the program looks like it is behaving normally since the results are what they are expected to be. This simply means that there is no cleaning done. This "feature", used by the wrong person, is a real concern. Therefore, reliable means have to be provided, in order to make sure that the machine has been thoroughly cleaned between two runs.

3 First Results

As we have been actively teaching, we have also seen the first results of the introduction of CUDA in real production codes. The first and easiest approach was to optimize library calls which can be described as 0D problems. Using a GPU can yield significant speedup in terms of total compute time since there are no dependencies between mesh cells. We have been able to measure a 3 fold speedup for the *whole run* when offloading equations of state to a GPU in a CFD code.

This is probably one of the ideal cases for GPU computing. In most cases, the code's architecture may not be suitable for GPU usage, *as is*. To extend a code towards GPU computing, the developer should follow the steps described thereafter.

First, the developer has to make sure that the code is *really* optimized. Then, he must locate what part of the data can remain on the GPU as long as possible. This is essential in order to avoid going back and forth from the main memory. Finally, the code should be checked for constructs, which might lead to troublesome or no parallel code segments on the GPU at all. In this case, it is best to rewrite some portions of the code.

```
outvec[i] = outvec[i] + matvals[j]*invec[indx[j]];
```

The example above features both indirect addressing and a reduction. Reduction is well known to be difficult to parallelize efficiently and indirect addressing can cause many bank conflicts if not considered with great care.

The migration to GPU computing is currently being investigated through the adaptation of a pollutant migration code. It uses a ray-tracing algorithm which is well suited for GPUs. From the first analysis, we already know that this port will require quite a number of man-months. This is due to the fact that the code structure must be reversed to cope with the GPU constraints.

4 Conclusion

This article has shown that a strategy for moving towards always more powerful computers has been implemented at CEA. Hardware has been installed in the CCRT for the users. Training sessions have been set to develop communities of programmers who will help spread the knowledge and help others think in terms of millions of tasks. The impact of this new technology on the production environment will be studied using the Grand Challenge strategy involving end users and sysadmins.

At the end of the year 2009, we will be able to have a stable and known machine for general production and a clearer view of the advantages of GPU computing. The CEA teams will then have a better idea of the path to multi petaflop followed by exascale computing.

Acknowledgement We wish to thank NVIDIA Corporation for kindly providing some of the illustrations.

Optical Interconnection Technology for the Next Generation Supercomputers

J. Sakai, A. Noda, M. Yamagishi, T. Ohtsuka, K. Sunaga, H. Sugita, H. Ono, K. Yashiki, H. Kouta

Abstract This article describes the new optical interconnection technology which would be indispensable to realize high throughput supercomputer. To achieve 20 Gbps signal transmission between two LSIs of the supercomputer, we have developed every necessary component of optical interconnect system such as vertical-cavity surface-emitting laser (VCSEL), PIN-photodiodes (PD), electrical transmission line, driver IC and receiver IC. Each device is designed to optimize high speed signal characteristics of throughout transmission system. With these components total transmission system between two LSIs is developed and 20-Gbps 100 m signal transmission is successfully demonstrated. This transmission system is applicable to actual supercomputer.

1 Introduction

In the development of the high throuput High Performance Computing (HPC) system such as supercomputers, there is an increasing demand for wide bandwidth between LSIs. To meet this requirement the signaling rate of LSIs is becoming faster and signal number is increasing. There is a prediction in the ITRS roadmaps that LSIs for HPC after 2010 are forecasted to have 20 Gbps 1000 signal inputs/outputs (I/Os) (Fig. 1) [1].

J. Sakai, A. Noda, H. Ono, K. Yashiki, H. Kouta
Nano Electronics Research Laboratories, NEC Corporation, 1753, Shimonumabe, Nakahara-ku, Kawasaki, Kanagawa, 211-8666, Japan, e-mail: j-sakai@bk.jp.nec.com

M. Yamagishi
Technology Development Division, NEC Engineering Ltd., Kawasaki, Kanagawa, Japan

T. Ohtsuka, K. Sunaga, H. Sugita
Device Platforms Research Laboratories, NEC Corporation, Sagamihara, Kanagawa, Japan

M. Resch et al. (eds.), *High Performance Computing on Vector Systems 2009*,
DOI 10.1007/978-3-642-03913-3_5

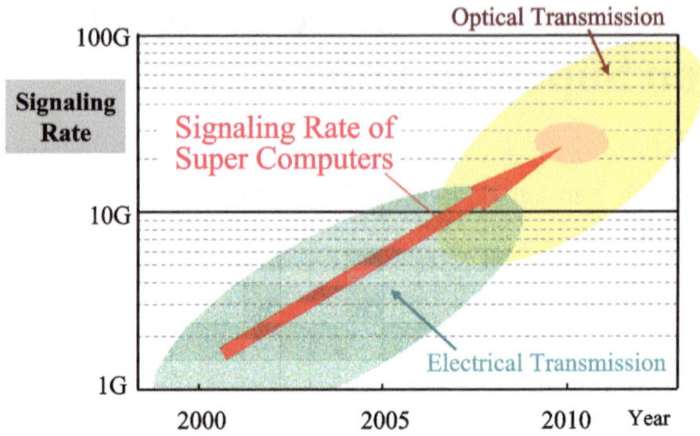

Fig. 1 ITRS roadmap of signaling rate of HPC

In the conventional system, it has been common to connect LSIs by electrical interconnection such as copper transmission lines. However, the transmission loss becomes large when signaling speed increases due to the dielectric loss, skin effect, and surface roughness. Therefore, high-speed signals faster than 20 Gbps can only propagate about 10 cm, although the transmission line length between HPC LSIs is approximately 50–100 cm. This indicates that electrical transmission faces the limitation of signaling rate and high throughput optical interconnection is required as major alternatives for upgrading interconnect performance.

Up to now optical interconnections above 10 Gbps have been demonstrated between optical ICs, such as driver and receiver ICs [2]. In actual computing systems over 20 Gbps/ch, not only the signal properties of optical interconnections but also those of electrical transmission lines between optical ICs and serializer/deserializer (SERDES) of LSIs have to be taken into account, because signal degradation and jitter accumulation in the electrical transmission path influence total signal integrity.

Previously, we designed an LSI unit for 1000 channel I/Os (Fig. 2) [3]. The optoelectronic modules were downsized and densely packed in order to be placed near the LSI, this helps to reduce transmission loss. And we developed high speed VCSEL and PIN-PD array that fit the high density LSI unit configuration [4].

We have since developed driver and receiver ICs suitable for over-20Gbps optical interconnections and for the LSI unit configuration described above. Each IC

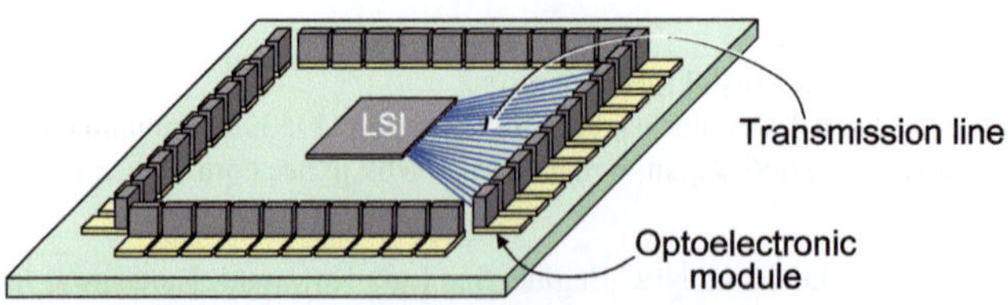

Fig. 2 Conceptual model of assembled LSI unit with optoelectronic modules for 1000 channels

is optimally tuned to the characteristics of the PD and VCSEL. 2R functions (re-amplification, re-shaping) of ICs, along with an equalizer of LSI, are designed to compensate for signal degradation in electrical lines. Further, we have fabricated LSI units, and have demonstrated 20 Gbps 100 m error-free transmission.

2 Components and Structure

The LSIs are fabricated in 65 nm CMOS technology, and each has two I/O sets for 20 Gbps SERDES. Transmitter (TX) circuit has a pseudo random bit sequence (PRBS) generator, and is equipped with a 5-taps feed forward equalizer (FFE). Receiver (RX) circuit has a linear equalizer and a built-in PRBS checker for on-chip bit-error-rate (BER) tests.

We have developed InGaAs quantum well (QW) VCSELs for their high speed and reliability [4]. Their bandwidth was 20 GHz. The PIN-PDs were InP-base back-illuminated mesa-type structure [4]. The 3-dB bandwidth was more than 20 GHz.

Driver and receiver ICs are fabricated in a SiGe-BiCMOS process to meet the requirements of high speed (above 20 Gbps) and low power consumption. In our circuit design, electrical equivalent circuit models of VCSEL and PD were introduced to optimize the combined characteristics of ICs and optical devices. IC designs for maximum eye openings at 20 Gbps were then determined. Each driver and receiver IC was equipped with an equalizer. The resulting power consumption values for the driver and receiver ICs in our design are 250 mW and 200 mW, respectively, half of what they would have been with a GaAs process.

Each TX and RX optoelectronic module was assembled on a $5.0 \times 5.0 \times 0.45$ mm alumina substrate with the ICs and the optical devices.

The motherboard is a build-up substrate, which is suitable for high-density wiring. The distance between each LSI and the optoelectronic module is 40 mm [3]. In order to reduce line loss, we employed a diagonal pair structure, which made it possible to widen the lines while still maintaining the same characteristic impedance [5]. The S21 at 10 GHz was 3.7 dB, which is 0.7 dB less than that in a conventional stacked pair.

The LSI unit was fabricated by mounting LSIs and TX and RX optoelectronic modules on a motherboard.

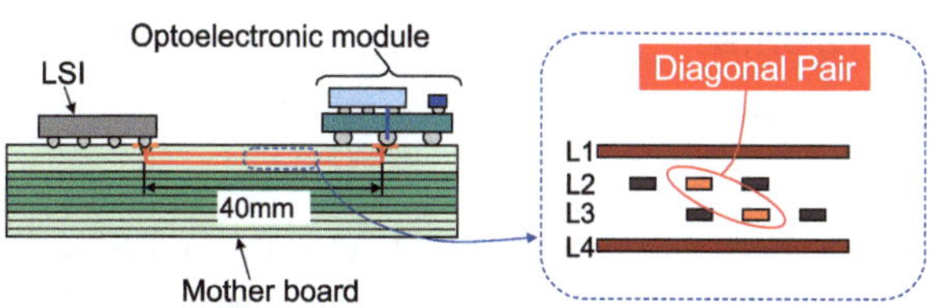

Fig. 3 Cross-sectional view of LSI unit

3 Performance

We measured the frequency domain transmission properties between the LSI and an optoelectronic module ("a" in Fig. 4). This property includes loss in the probe and cable used for the measurement, as well as transmission loss inside the LSI. With respect to this property, the FFE was optimized to maximize the eye opening at 20 Gbps ("b" in Fig. 4). Although total characteristics were relatively flat below 8 GHz, loss drastically increased over 12 GHz, which might be expected to lead to ISI ("c" in Fig. 4).

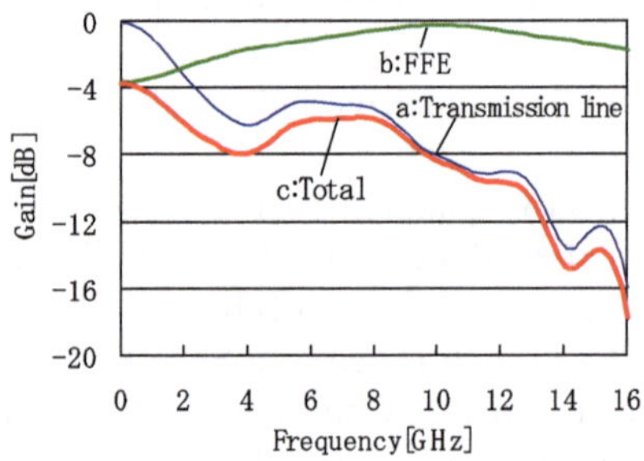

Fig. 4 Transmission properties between LSI and optoelectronic module

Figure 5 shows the small-signal frequency response of the driver and receiver ICs. The equalizer of each IC is designed to have a peak gain of roughly 11.5 GHz. The 3 dB bandwidths of the driver and receiver ICs are 18 GHz and 21 GHz, respectively.

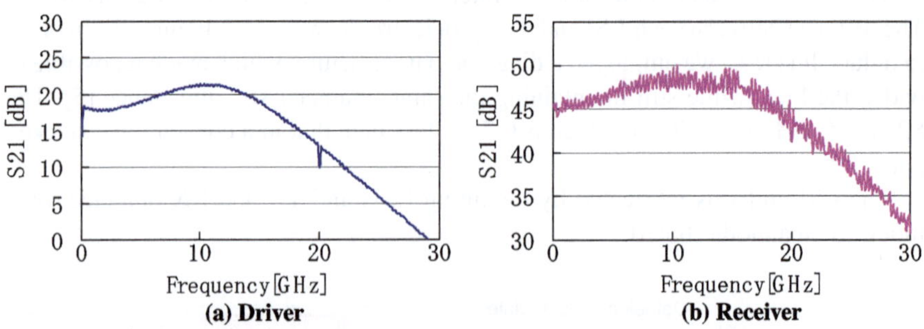

Fig. 5 Measured S-parameters of ICs

LSI-to-LSI signal transmission was carried out with two LSI units placed opposite each other. The VCSEL of one LSI unit was connected to the PIN-PD of the other with GI50 100 m multi-mode fiber. 20-Gbps 2^7-1 PRBS data, which corresponds to the 8 B/10 B that is generally used in actual systems, was transmitted from the LSI.

Figure 6 shows eye diagrams. The signal shown in Fig. 6(a) is equalized by the LSI and propagated through the substrate. It has a clear eye opening, but the 20–80% rise time is 22.2 ps, which suggests that degradation in the higher harmonics is nonnegligible. The driver IC amplifies this waveform by the above-mentioned frequency response and shapes it by a limiting function, and the VCSEL transmits it through 100 m MMF (Fig. 6(b)). Again the waveform is shaped by the receiver IC. As may be seen in the waveform in Fig. 6(c), the rise time of the input signal is only 19.5 ps, which means that the 2R functions of the driver and the receiver ICs have effectively shaped the waveform. An eye opening of 114 mV × 29 ps is obtained.

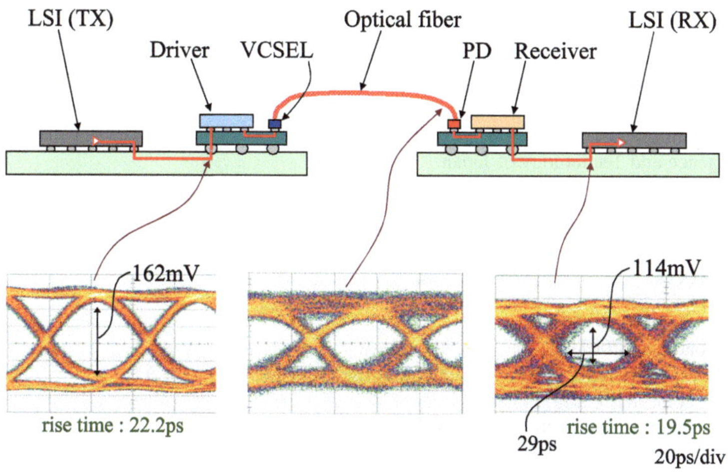

Fig. 6 20 Gbps eye diagrams for input signals to driver IC (**a**), PD (**b**), and receiver LSI (**c**)

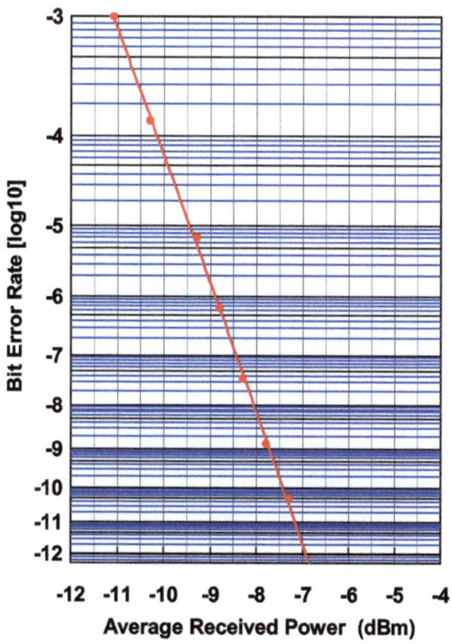

Fig. 7 Bit error rate for LSI to LSI transmission

Figure 7 shows BERs measured by the BIST circuit equipped on the RX LSI. Error-free (BER < 10–12) transmission was achieved at −6.5 dBm average optical power.

4 Conclusions

We have developed optoelectronic modules with driver and receiver ICs tuned for VCSELs, photodiode and electrical transmission property. 20-Gbps signal transmission was successfully demonstrated between two SERDES devices up to 100 m. These are applicable to actual high-end computing systems.

Acknowledgements A part of this work was supported by the Ministry of Education, Culture, Sports, Science and Technology of Japan.

References

1. ITRS Roadmap (2007)
2. L. Schares, et al., IEEE J. Sel. Top. Quantum Electron. **12**(5) 1032–1044
3. J. Sakai, et al., ICEP 2007, TC4-1.
4. K. Fukatsu, et al., 19th IPRM 2007, WeB304.
5. W. Fu, et al., 57th ECTC pp. 352–356 (2007)

HPC Architecture from Application Perspectives

Hiroshi Takahara

Abstract With the diversification and growing sophistication of high-performance computing applications, there are emerging requirements for the hardware and software infrastructure for peta-scale simulations and data handling. This paper addresses the key technological elements for realizing such a system from the architectural point of view. The author recognizes that the SIMD-based approach is getting pervasive as a core processor technology capable of providing a viable solution to the requirements for higher performance of applications and reduced performance-usability gap. This paper also gives an overview of the NEC approach in HPC with its innovative technological challenges toward future computing.

1 Introduction

The pursuit of faster supercomputers is fueled by our wish to advance the technological development and scientific findings. In this regard, a spectrum of emerging scientific applications are spearheading the evolution of high performance computing due to the demanding number crunching and intensive data handling with the increased model resolution and the sophistication of modeling techniques that encompass multidisciplinary approaches. On the other hand, the simulations in such areas as environment, energy, and nanotech fields that lie ahead require an enormous computing power to make them a reality. We still need 100 to 1000 times greater sustained performance with a challenging breakthrough, including the development of a new LSI device and high speed signal transmission. Figure 1 shows the general trend of peak performance gains of high performance computers during the past three decades. It has been tremendous with the average of more than 50%

Hiroshi Takahara
HPC Division, NEC Corporation, 1-10 Nisshin-cho, Fuchu, Tokyo 183-8501, Japan, e-mail: h-takahara@bc.jp.nec.com

M. Resch et al. (eds.), *High Performance Computing on Vector Systems 2009*,
DOI 10.1007/978-3-642-03913-3_6

per year for the entire period with its recent peak performance of around 1PFLOPS, including the NEC SX-9.

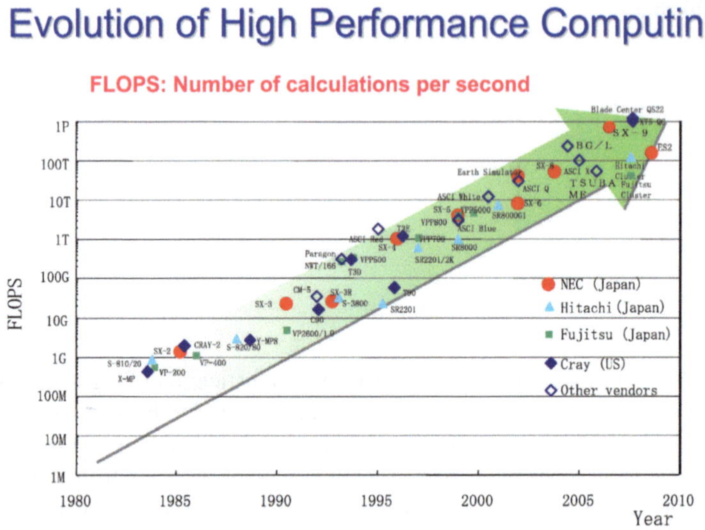

Fig. 1 Performance trend of HPC systems

With the increasing speed and sophistication of supercomputers, a third area of science, or computer simulation, has been emerging as a vehicle that can fully complement, or sometimes replace conventional approaches, such as laboratory experimentation/observation. For example, climate and weather modeling represents one of the Grand Challenges for high-end computing due to the enormous range of spatial and temporal scales that characterize the earth's processes. While emerging modeling techniques such as data assimilation, as well as coupled numerical models of the atmosphere, ocean, sea ice, land, and biosphere provide a feasible means for understanding the earth system across a wide spectrum of scales, they are also getting computationally expensive with the sophistication of algorithms. Improving forecasts of severe events such as hurricanes and tornadoes has long been a target of weather research in pursuing the reduced loss of life and property; however, such high-resolution local forecasting requires demanding HPC resources.

NEC has been conducting research and development toward the era of peta-scale computing or beyond that. Extensive efforts are needed for various technologies ranging from CPU architecture, LSI packaging, power consumption, and cooling method to massive parallelism and job scheduling, and even to the handling of a volume of data resulting from intensive number crunching. These elements are particularly crucial in realizing high sustained performance on real scientific and engineering applications. In terms of hardware designing, the key technologies include high-speed, highly integrated LSI designing and fabrication, fast interconnection (optical device, optical link etc.), high-density packaging, and fast memory device with lower power consumption. In addition, advanced software technologies are also an integral part of future high performance computing, such as the manipulation of a large number of jobs and ultra-massive parallelism.

This paper outlines the architectural trend from the view point of the characteristics of core CPUs and the performance of application software with its future perspectives.

2 Trend of CPU Performance

Figure 2 shows the trend of single-CPU core performance during the past fifteen years. Apparently, there is a widening gap between the vector processors (such as the NEC SX Series) and scalar processors with respect to single-core performance especially after 2004, implying the necessity of drastic technological breakthroughs.

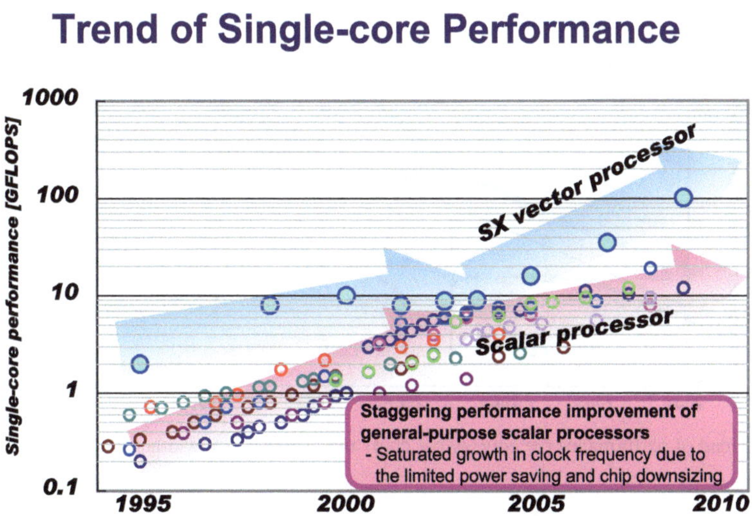

Fig. 2 Trend of single-core performance

Most of the current core processors that enable high performance adopt the general-purpose microprocessor technology with the scalar CPU architecture. While these processors are getting commoditized, the enhancement of the clock speed of the CPU has stagnated due to the increased power consumption (leak current) and heat generation, as well as saturated chip downsizing. In order to offset such disadvantages, the multi-core technologies are gaining momentum in enhancing the peak computing performance. The single-core (single-chip) performance of the SX-9 particularly surpasses those of commodity-based cores by a factor of almost 10, thus allowing the user to obtain high performance much easily without spending much effort for massively parallel programming.

At the same time, it is getting an ever more fundamental issue to ensure high sustained performance on real scientific application programs. Particularly after 2003, the gap of the system performance on the Linpack score and the actual performance on application programs is getting obvious [1].

Figure 3 indicates the trend of the memory bandwidth for a core and a whole chip made up of multi-cores together with that of single-core and shingle-chip performance. The single-core performance of commodity-based chips has almost been saturated, which hampers the stable growth of CPU performance. The trend of the single-core memory bandwidth is even on a decline.

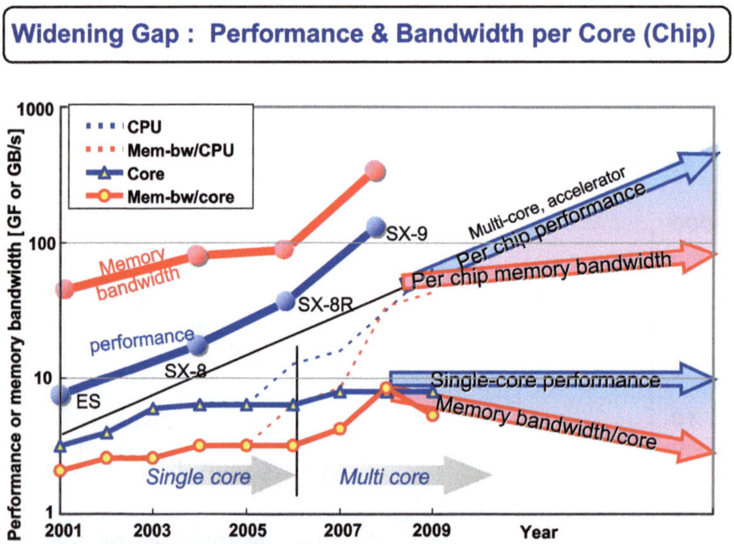

Fig. 3 Trend of performance per core (or per chip) and memory bandwidth

The reasons behind such a performance trend are the rising power consumption and the limitation to the cooling capacity of LSIs. Another factor also encouraging this trend is the fact that the single-core (or single-CPU) performance tends to be saturated without increasing in proportion to the number of devices mountable on a substrate. Inevitably, many computer vendors are heavily relying on multi-core technologies in order to enhance the capabilities of CPUs, which in turn tends to hamper ease of programming for high sustained performance. The user often has to cope with increasingly complicated programming for maintaining cache coherence or alleviating cache contention.

Anyway, the memory bandwidth is not necessarily commensurate with the growing performance of a CPU chip. On the other hand, the single CPU performance of the NEC SX Series has been increased so far toward the SX-9. Its memory bandwidth has also been enhanced, enabling a rather balanced system performance in contrast to commodity-based scalar chips.

3 Architectural Challenges

In pursuing high performance computing, there is a growing concern over the increased energy consumption and the required floor space together with conventional requirements for high sustained performance, ease of programming and system operation, inheritance of application-software assets, and system scalability.

Figure 4 indicates such technical issues that would need a novel approach.

The first challenge is the deceleration of the reduction of power consumption. As a recent trend, the performance increase per power consumption has been limited to a factor of only two for five years. The second limitation to the performance enhancement is the upper limit to the power equipment facility and the floor space. Another major challenge comes from the limited power efficiency due to the staggering improvement in device miniaturization. Furthermore, we are faced with the performance ceiling resulting from the increasingly challenging architectural simplification of CPU cores and multi-core technologies, thus bringing about the slowdown in performance improvement.

Some performance improvement mechanisms are essential for bridging the gap between the expected performance and the current performance estimation toward realizing peta-scale computing, such as the adoption of a performance acceleration mechanism, for example.

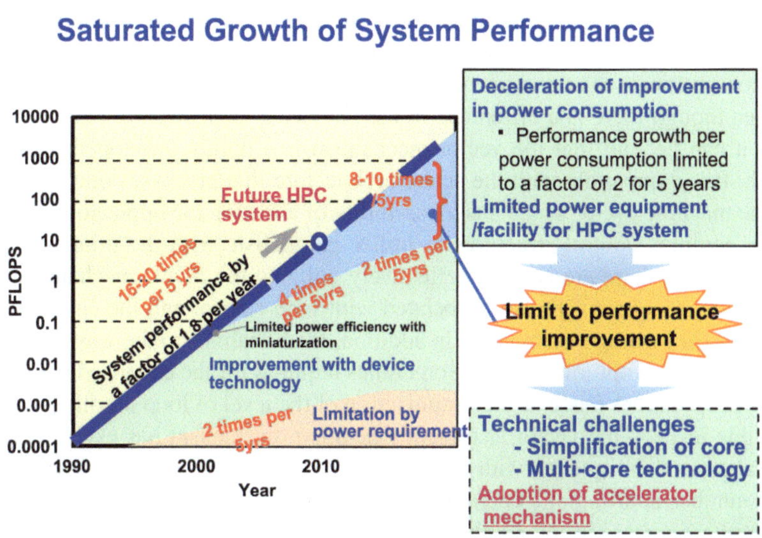

Fig. 4 Architectural issues in performance enhancement

4 SIMD-based Approaches

Under the current situation where the improvement of clock speed is slowing down, it is necessary to enhance the instruction-level parallelism within a scalar core processor for the continuous improvement of the performance of a scalar core processor together with the increase in the number of arithmetic operation units. While this is a possible approach toward peta-scale computing, the obtained performance gain is not necessarily sufficient with respect to the invested hardware amount required for extracting the instructions that can be run concurrently.

Furthermore, complex control mechanisms and massive hardware resources are necessary for realizing a high sustained performance with the current scalar architecture, which tends to cause a reduced performance-to-power consumption ratio. To overcome such an issue, a certain accelerator mechanism is expected to leverage the performance per energy consumption.

As a performance acceleration mechanism for high performance computing, there are two means; one is the Single Instruction Multiple Data stream (SIMD) arithmetic operation mechanism that is built into some scalar processors mainly for multimedia data processing and the other the vector operation mechanism which has been adopted for conventional supercomputers. The SIMD and vector architectures both attempt to capture data-level parallelism by getting a single instruction to perform multiple operations concurrently. In both cases, the instructions specify a single type of operation and define individual data elements to be independent [2]. Here the former SIMD mechanism is aimed at general-purpose use and operates several arithmetic instructions at the same time per one machine instruction (usually per one machine cycle). On the other hand, the major trend of the latter vector arithmetic mechanism is represented by multiple vector pipelines that process several tens to hundreds of arithmetic operations concurrently within several machine cycles. This allows the user to easily gain higher sustained performance in comparison with conventional scalar architectures [3].

Even though there is a difference in the number of concurrently issued instructions, it can be said that the vector mechanism is a highly developed SIMD approach. It is compatible with the scalar architecture that processes one instruction per one machine cycle, which has a possibility of allowing the application program having complex control structures to utilize parallelism with the minimum grain size. However, there are issues in effectively utilizing the SIMD mechanism, such as the mitigation of overheads associated with loop control and the calculation of addresses for loading/storing data. In addition, modifications to the way of application performance optimization are sometimes required for the enhanced parallelism in SIMD operations (such as the optimization of the level of loop unrolling).

In this respect, the vector arithmetic operation mechanism has excellent architectural expandability even with the same SIMD technology. This is because of a larger number of arithmetic operations handled by one instruction, which can facilitate further increase in the number of arithmetic operations by the use of multiple pipelines. In addition, no significant efforts are needed for restructuring application programs when enhancing parallelism. Anyway, vector operations provide an easy way to express the concurrency needed to take advantages of the memory bandwidth with more tolerance about the memory systems' latency than scalar designs. Vector operations can also help mitigate bottlenecks in memory, power, and performance for data-parallel applications such as scientific and multimedia codes.

As shown in Figure 5, the SIMD approach, which is also a basis of the vector arithmetic operation mechanism, is adopted for a wide range of architectures ranging from general-purpose processors and add-on co-processors to heterogeneous multi-core CPUs (such as the Cell architecture), signaling a new trend in computer architecture for high performance computing.

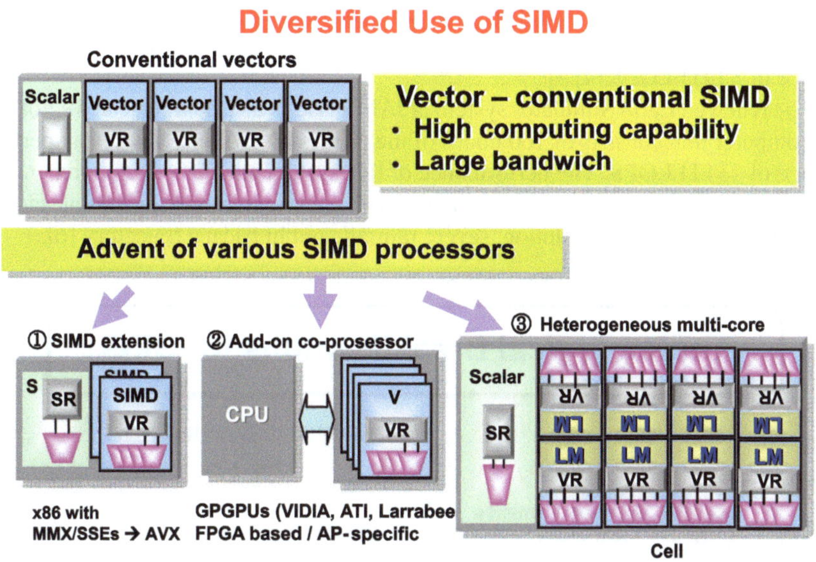

Fig. 5 SIMD architecture utilized for diversified processors

Together with the processor architecture, one of the key technologies indispensable to achieving higher system efficiency is the interconnection between processors. NEC has developed an innovative optical interconnection technology that achieves a 20 Gbps signal transmission rate between two LSIs of the high performance computer under the support of the Ministry of Education, Culture, Sports, Science and Technology of Japan. It includes the development of necessary components of an optical interconnection system, such as vertical-cavity surface-emitting laser and PIN-photodiodes. These devices are aimed at optimizing high speed signal characteristics throughout the transmission system, allowing two LSIs placed at a distance of 100 meters to realize a 20 Gbps signal transmission rate. This is a technological breakthrough surpassing the existing electrical transmission in terms of communication speed, which is expected to become an integral part of the interconnection technologies needed in 2010 or later.

With respect to the system architecture capitalizing on the SIMD technology, NEC has been offering the SX Series parallel vector computers for scientific and engineering applications. Figure 6 summarizes some of the core technologies built into the SX Series for enhanced performance and ease of use, which range from the LSI technology and the high-speed interface and high-density packaging technologies to the advanced cooling technology.

The SX Series has the advanced architecture of large-scale shared memory, high-speed data transfer between the CPU and the memory, and an ultra high speed network connecting nodes. Its latest model SX-9 boasts the world fastest single CPU performance (102.4GFLOPS) currently available in the market, making up one node system of a 1.6TFLOPS peak performance with 16CPUs. The system can be configurable up to the maximum of 512 nodes having its theoretical peak vector performance of 839TFLOPS [4, 5].

The renewed Earth Simulator system at JAMSTEC is also a large-scale vector supercomputer that consists of 160 nodes of the NEC's SX-9/E and has a peak performance of 131TFLOPS. The performance of this system enables the comprehensive understanding of the complex phenomena with greater accuracy, including global warming, air and ocean pollution, severe rainfall, and hurricane trajectory [6].

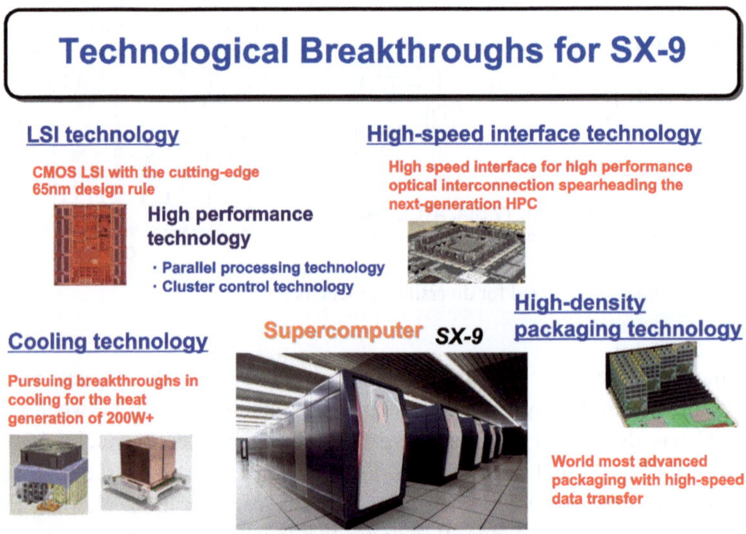

Fig. 6 Technologies for the SX-9

5 Conclusions

Looking forward, NEC will continue to strengthen the development of advanced high performance systems while also providing customers with optimal systems

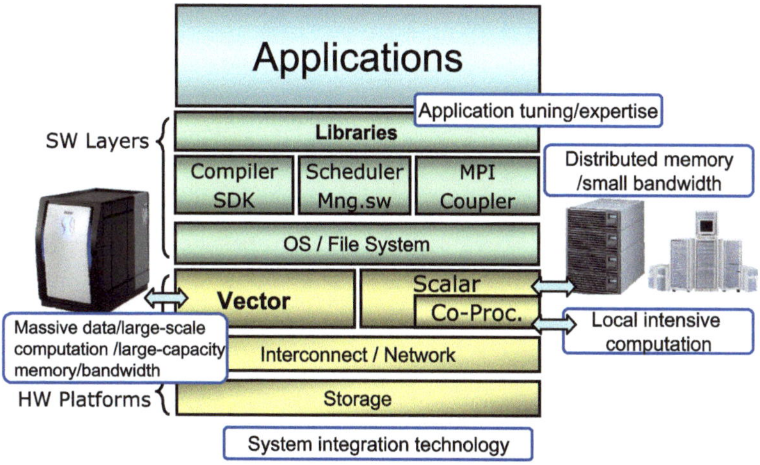

Fig. 7 Components of the NEC HPC systems

through scalar supercomputers, PC cluster systems, and GPGPUs (co-processors) depending on the requirements and the nature of application codes (Figure 7). Improved computing capabilities are central to the development of scientific research and engineering, such as the challenging global environmental issues and nano-material designing. NEC will advance the SIMD-based approach on the basis of its experience and expertise and will promote the technological evolution for future high performance computing toward peta-scale simulations.

References

1. HECRTF: Federal plan for high-end computing: report of the high-end computing revitalization task force (HECRTF). http://www.nitrd.gov/pubs/2004_hecrtf/20040702_hecrtf.pdf (2004)
2. Gebis, J., Patterson, D., Embracing and extending 20th-century instruction set architectures. Computer **40**(4) (2007) 68–75 (published by the IEEE Computer Society)
3. Oliker, L., et al., Leading computational methods on scalar and vector HEC platforms. IEEE/ACM SC2005 (2005)
4. NEC Technical Journal (Supercomputer SX-9/Special Issue), vol. 3(4). http://www.nec.co.jp/techrep/en/journal/g08/n04/g0804mo.html#name3-1 (2008)
5. NEC SX-9 Supercomputer. http://www.nec.de/products-services/high-performance-computing/index.html (2008)
6. System overview of the new Earth Simulator (JAMSTEC). http://www.jamstec.go.jp/es/en/system/system.html (2009)

II
Strategies

A Language for Fortran Source to Source Transformation

Uwe Küster

Abstract Based on experiences with imperfect compiler optimization we try to develop a tool which transforms source to source code by user defined transformations. The language provides mechanisms and constructs which facilitate the parsing of codes, the collection of the necessary information, the transformation of the code segments and the definition of the transformed code including all features to get a well defined transformed program. We shortly describe the language and give some examples.

1 Compiler

Modern computer architectures may deliver very high floating point performance. But they do this on the base of deep implicit and explicit parallelism. And they suffer on restricted memory bandwidth and large memory latencies, which limit the effective performance unless caches, local memories and register can be reused for providing the necessary data and the unavoidable latencies may be hidden by prefetching the data. Modern compilers have to produce code which is effectively using the hardware features of the architecture. There are some well known transformations enabling performance as loop unrolling loop fusion, loop blocking, inlining. But a compiler as an automatic tool can not understand the user intentions. The compiler is forced to generate accurate code under all and very unlikely circumstances. It cannot estimate loop iterations counts, does not see effective bounds of algorithmically relevant parameters, can not recognize injective index arrays or other memory access patterns, can not distinguish overlapping and nonoverlapping pointers. Run time tests for analysis of the actual situation, for example the calculations of the address range of an array which will be covered within a loop, would consume computing time and would be contraproductive in some cases. Complementing the

Uwe Küster
HLRS, Nobelstraße 19, 70569 Stuttgart, Germany, e-mail: kuester@hlrs.de

M. Resch et al. (eds.), *High Performance Computing on Vector Systems 2009*,
DOI 10.1007/978-3-642-03913-3_7

languages by constructs ensuring better optimization and parallelization seems to be a way. The array syntax of Fortran could serve as an example but has performance inhibiting semantics as the assumption that the complete right hand side of an array expression has to be evaluated before assigning the data to the left hand side. More complicated data structures could be language primitives as linked lists with unique elements or explicit trees without cross references. Directives are a generalization of languages enabling the programmer to show semantics of the coded algorithm. Directives expressing that a loop has no inner dependencies, limiting the loop iteration count, enabling prefetching, triggering loop unrolling and loop collapsing may be essential for code optimization. They enable inlining, influence alignments and vectorization and may show the frequency of branch. But syntax and meaning of these directives are not portable between compilers and machines except for the standardized parallelization paradigm OpenMP as an very important example.

Most directives are bound to loops. They influence directly the control constructs of a loop. They do not disclose properties of arrays in a larger code segment or in a data structure. If the programmer knows about some special features of his algorithm it will not be possible for him to express this in a way enabling new transformations. He has to implement the transformations himself. These may generate confusing code with many repetitions which is difficult to maintain and may even be contraproductive on other hardware architectures.

2 Self Defined Transformations

We think that a mechanism to define annotating directives which initiate problem dependent code transformations would helpful for the generation of maintainable and fast code. The programmer would implement some of these directives at some appropriate positions. The transformation program would do the transformations at the designated location. The transformed program which is still readable source code will be compiled in the next step. The original code remains unchanged except the additional directives.

For this purpose we have developed a transformation language which allows for parsing new user defined directives within the source code and for formulation of the appropriate transformations. The transformation language has constructs to parse the source code and to transform it in optimized code. The supported source code language is Fortran. The technology could be also applied to other languages. In the following we will describe the transformation language and give some examples.

3 The Transformation Language

The transformation language allows for an abstraction of transformation rules. The language defined in a way that the written transformation programs may transform source code to source code by the implemented rules. We will describe the elements of the language. The language has transformation variables, the transformation control constructs and a library of the intrinsic procedures.

3.1 Transformation Variables

The value of a transformation variable can accessed by preceeding its name by "@" similar to scripting languages. Transformation variables can be

- strings,
- integers,
- logical values,
- tokens of the Fortran code. The Fortran source code is represented by a linked list of its tokens,
- Segments, which are the interval between two connected tokens, for example a statement or an assignment,
- Fortran names (names of variables, procedures, modules, ...),
- set of Fortran names and so called,
- designs which are themselves collections of strings, integers, logicals, segments, Fortran variables and designs. Each of these parts is described by a selection parameter which can be used to access the part. Wild cards in the selection parameters are allowed.

The transformation variables have not to be declared, they get their type by an assignment or as resulting value of an intrinsic procedure. The Fortran code is not represented by a tree but by a linked list, because handling of parts of segments is simpler.

3.2 Transformation Constructs

In the transformation modus the control constructs and the transformation variables are accessible. The control constructs and intrinsics support the formulation of the transformation programs.

- The loop control construct is
  §do(it_ var = loop_ control_ expression)... §enddo(it_ var)
  The do loop may include cycle(it_ var) to get the next iteration or exit(outer_it_ var) to leave the loop. Both include the name of an iteration variable. This allows to leave also an outer loop. The form of the loop_ control_ expression allows for an appropriate scanning of sets. It may be consist on a start and an end integer or a segment with an statement or token attribute. In the first case all statements of the segment are tested and in the second case all tokens. The loop_ control_

expression may also be a design with a supplemental expression of selection parameters.

- The branching construct is
  §if(⟨logical_ expression⟩) §then(if_ name)
  ...
  §elseif(if_ name)
  ...
  §else(if_ name)
  ...
  §endif(if_ name)
  with the obvious meaning. The construct names in parenthesis ensure a proper nesting.
- §set(variable = set_ expression) to define and set transformation variables. expression is similar to the loop_ control_ expression in a do loop. It may also contain
- §evaluate(expression) which evaluates an arithmetic or string expression.
- §get(@variable, selection_parameter,movement,result) to extract the content from transformation variables of different data types and for the definition of new transformation variables; movement can be composed expressions simplifying the orientation in constructs as Fortran do loop, select case constructs or if then else endif constructs.
- §add_to(@design,para,@entity) stores a token or a segment or other design parts from entity in design. The stored content can be found later by using the parameter para as selection_parameter in §set(...) or §get(...) or §do(...).
- §print(string, @variable) prints transformation variables of all types in variable.

3.3 *Self Defined Procedures in the Transformation Code*

The transformation language knows self defined procedures which are called via §exe(procedure_ name, list_ of_ procedure_ arguments). All variables in the procedure are global except these which are declared as local.

3.4 *Intrinsic Procedures*

The transformation language contains a large variety of different intrinsic functions. Some of these may detect and parse Fortran constructs as

- §is_ function_ or_ array(...),
- §is_ assignment(...),
- §is_ do_ enddo(...),

If successful they return a design containing the important parameters of the constructs which can be later used for transformations.

- §parse_ all(...)
  parses a construct found by help of parsing primitives and also delivers the parameters in design variables. The construct may be a Fortran statement or also a self defined directive.

There are intrinsics for maneuvring in the Fortran text like

- §next_active_statement(@actual_token,new_token),
- §last_statement(@actual_token,new_token).

Other intrinsics handle the transformation variables of the transformation language itself as

- §var_decl(@variable,@to_variable_set,declaration_list)
  incorporates the intended Fortran variable in the to_variable_set and defines declarations attributes
- §add_declarations(@insert_token,@varset)
  adds all declarations collected in varset at the point insert_token.
- §find_variable_in_varset(...)
  finds specified Fortran variables in a segment in the Fortran text.

3.5 Parsing Primitives in Parsing Mode

The parsing primitives have a different syntax and serve for parsing constructs of the Fortran code. They begin all with a #-sign. They can be combined with the transformation constructs. As examples can be understood

- #name parses a Fortran name for variables, arrays, derived types, procedures, e.t.c.
- #bos and #eos parse the begin and the end of an instruction
- #arithmetic_expression parses an arithmetic expression
- #function_or_array parses a function call or an array evaluation
- #label parses a label
- #assignment parses an assignment
- # parses a blank, and #, parses a comma.

A name like #?var_name is used to store the result of the subsequent primitive or group of primitives as entity which may be accessed later by the parameter *var_name* in the resulting design. The following expression can parse the begin of a Fortran do loop as

```
do  iter = iter_start, iter_end, iter_diff
```

Remark the variables storing the different parts as segments

```
#?begin_construct do
#?iterator #name
=
#?iteration_start #arithmetic_expression
#?iteration_end #arithmetic_expression
#?increment_switch #{ , #?iteration_increment
  #arithmetic_expression #}
```

Primitives and constructs may be combined together via three types of parentheses to other constructs. The parenthesis may be nested. All these primitives and the different parenthesis can be preceeded by parsing variable names of the type #?var_name.

- #(... #) groups entities. A preceeding parsing variable name stores the segment of the whole group if the parsing is successful.
- #{ ... #} are handled as true if the enclosed sequence appears at least once. A preceeding parsing variable name stores the number of successful repetitions of the whole construct if the parsing is successful or 0 otherwise.
- #[... #] are handled as true if one part of the enclosed sequence appears. A preceeding parsing variable name stores the segment of the first successful entity.

If the parsing of the combined sequence is successful the parsing step delivers a resulting design of all segments and repetition numbers which are accessible via the names of the parsing variables.

4 Examples

The following example is the result of a unrolling program written in the transformation language. Additionally we see a sequence of the transformation code used for the parsing of the unrolling directive. The syntax of the directive and the transformation allow for a even more sophisticated unrolling techniques as we see it in usual commercial compilers. Remark also the handling of the source code instructions between the nested loops.

```
!directive begin unroll j by 2
 do j = i+1, jmax
   ss = 0.0d0
   do k = i+1, kmax
     ss = ss + aa(j,k) * aa(i,k)
   enddo
   ff=ss/hh
   do k = i+1, kmax
     aa(j,k) = aa(j,k) + ff * aa(i,k)
   enddo
 enddo
!directive end unroll
```

This code will be transformed to the outer unrolled code

```
   treated!!directive begin unroll j by 2
     j_start=i+1
     j_d=mod(jmax-j_start+1,2) ! case_number
     j_shift=j_start+j_d
     j=j_start ! first it_variable
    select case(j_d)
    case(0)
    case(1)
      ss_j0 = 0.0d0
      do k = i+1, kmax
        ss_j0 = ss_j0 + aa(j+0,k) * aa(i,k)
      enddo
      ff_j0=ss_j0/hh
      do k = i+1, kmax
        aa(j+0,k) = aa(j+0,k) + ff_j0 * aa(i,k)
      enddo
     end select
     do j=j_shift,jmax,2
      ss_j0 = 0.0d0
      ss_j1 = 0.0d0
      do k = i+1, kmax
        ss_j0 = ss_j0 + aa(j+0,k) * aa(i,k)
        ss_j1 = ss_j1 + aa(j+1,k) * aa(i,k)
      enddo
      ff_j0=ss_j0/hh
      ff_j1=ss_j1/hh
      do k = i+1, kmax
        aa(j+0,k) = aa(j+0,k) + ff_j0 * aa(i,k)
        aa(j+1,k) = aa(j+1,k) + ff_j1 * aa(i,k)
      enddo
     enddo !new
   treated!!directive end unroll
```

For parsing the directive in the example we use the following sequence in the transformation program.

```
§set(str1=' #bos ! @DIRECTIVE begin unroll ')
§set(str2=' #?induction_variable #name ')
§set(str3=' #?keyword #name ')
§set(str4=' #{ #switch_comma #?number #integer #} ')
§set(str5=' #?only_switch #{ only #} #eos ')
...
§if( §parse_all(@statement_token,parse_para,
@str1,@str2,@str3,@str4,@str5) )
```

```
§then(if_parse)
#* we extract the induction variable from the directive
§if(§get(@parse_para,induction_variable,string,
        induction_var))
§then(if_ind)
§print(++++unroll directive++++-,@statement_token)
§endif(if_ind)
§if(§get(@parse_para,keyword,string,unroll_case))
§then()
§print(got the unroll case ,@unroll_case)
§endif()
...
§endif(if_parse)
```

5 Concluding Remarks

We are developing a tool which helps for performance enabling by restructuring the source code. The tool is programmable and can express special source to source transformations which make sense for a broad range of applications or also only for a very specific user code. The tool is not restricted to situations which are completely transparent to a compiler which is forced to generate correct code under all circumstances. The produced code is still accessible to the user and could be analysed for correctness. Because the transformation program is open to the user it could be adjusted in the right way. The transformation language is not easy but relatively simple to handle. Up to now we have implemented some transformations which are useful to get higher performance on vector systems.

The Rose Compiler project (http://www.rosecompiler.org) has similar intentions. Rose provides also mechanisms for building source to source transformations. It does this on a sophisticated intermediate inner representation by an abstract syntax tree. On that base automatic transformation programs can be build. The approach is more general but also more difficult to understand and to handle.

The SX-Linux Project: A Progress Report

Erich Focht, Jaka Močnik, Fredrik Unger, Danny Sternkopf, Marko Novak,
Thomas Grossmann

Abstract The increasing appeal of hybrid computer architectures built of components specialized for different purposes and types of programs motivated this project that is aimed at integrating the NEC SX architecture more easily with hardware running the Linux operating system. Users of the NEC SX vector computers should be presented a Linux-like environment that simplifies the usage and porting of code developed on Linux workstations, and allows the use of parallel filesystems like Lustre directly from applications running on the SX nodes. The project is a collaboration of HLRS and NEC EHPCTC Stuttgart and has started in the first half of the year 2008. This paper reports the progress after one year of work with a team of up to 6 people: a working open source toolchain with binutils and the C compiler of the GNU Compiler collection ported to the SX, a simple free C library usable for testing, progress on the GNU C++ compiler, an I/O forwarding infrastructure that allows SX programs to do I/O to files visible on Linux servers, and a running light weight kernel derived from Linux code, that is a sane basis for developing a high performance specialized compute node kernel for the NEC SX machines.

Erich Focht, Fredrik Unger, Danny Sternkopf
NEC HPC Europe, Hessbrühlstr. 21b, 70565 Stuttgart, Germany, e-mail: efocht@hpce.nec.com, funger@hpce.nec.com, dsternkopf@hpce.nec.com

Jaka Močnik, Marko Novak
XLAB Research, Pot za Brdom 100, 1000 Ljubljana, Slovenia, e-mail: jaka.mocnik@xlab.si, marko.novak@xlab.si

Thomas Grossmann
High Performance Computing Center Stuttgart (HLRS), Nobelstr. 19, 70569 Stuttgart, Germany, e-mail: grossmann@hlrs.de

M. Resch et al. (eds.), *High Performance Computing on Vector Systems 2009*,
DOI 10.1007/978-3-642-03913-3_8 © Springer-Verlag Berlin Heidelberg 2010

1 Introduction

The evolution of hardware used in high performance computing (HPC) systems is following a trend away from general purpose CPUs to more and more specialized hardware: vector CPUs, GPUs, FPGAs, hybrid processors like the Cell. At the same time general purpose processors extended their SIMD units aiming at a more efficient use of their arithmetic pipelines for data parallel code. The variety of hardware made available to application development becomes harder to deal with, and it appears natural that some applications are tuned for running extremely well on certain hardware. General purpose HPC data centers like HLRS thus offer a broad choice of computers and infrastructure, trying to suite the various requirements of more or less specialized applications. Integration of various computer systems naturally leads to hybrid systems, for example coupled scalar-vector systems, as envisioned at HLRS and pursued within the TERAFLOP Workbench project [1].

The drive behind the project, creating an open platform for the SX is to be able to keep up with the development in the Open Source world, simplifying contributions and creating new models of usage for the vector computer. The need to integrate the SX vector computer into compute centers where Linux already is prevalent is also a force. Another important angle is the possibility to learn from system software development and draw parallels with other architectures and improve their system software.

The paper is structured as follows: Sect. 2 describes and motivates the project paths. Next, Sect. 3 describes the progress up to this point detailing the software developed so far: The GNU Toolchain in Sect. 3.1, User space and I/O Forwarding in Sect. 3.2 and Kernel in Sect. 3.3. Finally we conclude with an outlook.

2 Project Paths

According to the Wikipedia definition "Linux is a general term referring to a UNIX-like operating system based on the Linux-kernel". What exactly is meant by the word "Linux" is depending very much on the person asked and its area of interest. A kernel developer is focused on the Linux kernel, while application programmers rely on the user APIs provided by the common libraries, most notably the *GNU libc*. Administrators, on the other hand, use many small GNU-UNIX tools for managing Linux computers, and expect to be able to use certain features like particular filesystems.

Since neither a Linux kernel, nor a GNU compiler, toolchain or glibc exist for the NEC SX, the development targets cover a very broad area. Figure 1 sketches the project paths in a way similar to mind maps.

The left-most path is dedicated to developing an open and free GNU toolchain for the SX CPU. Besides opening up the SX architecture and making it easier to develop and build Linux software for it, binutils and gcc are prerequisites for any Linux-related kernel work. While vectorization is not absolutely necessary for a compiler aimed at system software development, it constitutes a possible side-project.

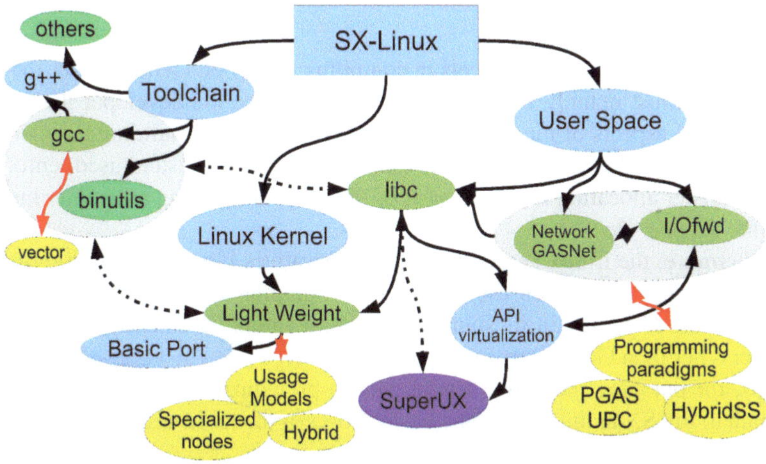

Fig. 1 Project paths mindmap. Advanced components are displayed in green, possible side-projects are yellow

The right-most path in Fig. 1 represents the user-space approach to providing a Linux-view on the SX. It aims at developing

- An I/O forwarding mechanism on top of a generic and adaptable network layer, that can be embedded into a libc under Super-UX and Linux. The I/O forwarding should provide a practical way to use filesystems mounted on Linux fileservers from the SX compute nodes.
- A free C library usable on top of both kinds of kernels: Super-UX and Linux.
- A system call virtualization layer below the C library which allows flexible adaptation of I/O forwarding, but also the use of a Linux-like glibc on top of Super-UX.

Possible side projects for the user-space path are the development of a hybrid programming model building on top of the I/O forwarding mechanisms and work on PGAS languages that would use the I/O forwarding network stack.

The middle path in Fig. 1 is aiming at the development of a kernel for the SX that allows running Linux user space programs. The investigation of the difficulties expected on this path showed that a full fledged Linux kernel with all features needed in a data center probably cannot be developed in the time frame of the project. Instead of porting directly a basic version of the Linux kernel we decided to analyze the difficulties in a simpler framework: a light weight kernel (LWK) based on Linux [2]. The rationale was to find out early whether there are any insurmountable problems for this project path. If not then the work done on the LWK can be either moved into a basic Linux port or extended for an optimized light weight kernel specialized for SX compute nodes.

3 Progress and Status

3.1 The GNU Toolchain

While the native SX toolchain excels in compiling highly efficient code for the vector CPUs, it, most notably the C compiler, can not give the developer the control over low-level aspects of the output code that is required for building an operating system kernel for the SX architecture: limiting the set of registers used, enforcing manual register allocation for some parts of the code, embedding assembler instructions into C code, etc.

Furthermore, the Linux kernel [3] and its derivatives like the Kitten lightweight kernel [2] heavily depend on use of the GNU Compiler Collection (GCC) [4] extensions to the C syntax in its source: building a Linux (or Kitten) kernel with a compiler other than GCC proves prohibitively hard.

Thus, for pursuing the native Linux path, a prerequisite was to port the GNU toolchain to the SX architecture. The basic subset of the toolchain consists of:

- *binutils* [5], comprising the basic tools for inspecting and manipulating object and executable files, such as *objcopy, objdump, ar, ranlib and strip*, the GNU assembler *gas*, and the GNU linker *ld*,
- the *GNU Compiler Collection, gcc*, that provides frontend compilers for a number of languages (C, C++, Objective C, Fortran, Ada) and a backend for a variety of processor architectures.

The emphasis was put on the C compiler, and C++ compiler was set as a secondary goal. Once again we note that the goal of the port was not to produce a competitive vectorizing compiler, but a basic C compiler for scalar code that would allow us to build low-level system software, such as an operating system kernel.

3.1.1 Binutils

The binutils tools had to be ported first, and binutils 2.19 (current stable version as of this writing) were used as the base for the port.

The porting first required adding support for the SX flavour of the COFF binary format, which differs from ordinary COFF in sizes of some header fields, and a number of SX-specific flags (floating point type, memory layout etc.).

Binutils handle different binary formats (predominantly variations of COFF and ELF binary formats) through the BFD (Binary File Descriptor) library, which needed extending in order to support the SX COFF specific differences.

Then, the SX CPU opcode and mnemonic descriptions were added, which serve as the basis for assembling and disassembling the code. SX-4 to SX-8 instruction sets are currently supported (including all vector instructions as well as instruction dealing with special registers), implementation of SX-9 support is pending.

With these two patches in place, common tools could be used to inspect and modify SX COFF binary files. Most notably, this produced a working SX disassembler (as part of the *objdump* tool), which the native SX toolchain lacked.

The GNU assembler port required a parser for the SX assembler syntax, and support for the SX specific pseudo operations (such as *using* etc.) and semantics (implicit base addressing via base register set with the *using* pseudo operation etc.).

A few extensions, most notably support for weak symbol semantics, not present in the native SX toolchain were also added to the assembler.

The GNU linker required no explicit porting, except for writing the link scripts that arrange the code and data into a layout expected by the Super-UX operating system. The port diverged into two targets at this point:

- *sx?-nec-superux* target with link scripts that produces an executable with a memory layout expected by Super-UX, and provide a number of symbols that are required by the Super-UX C and C++ run-time libraries. The values of these symbols were deduced from the code produced by the native SX toolchain due to lack of documentation on them.
- *sx?-nec-linux* target that omits the above mentioned Super-UX specific symbols, and is used for building the experimental Linux-like kernels for the SX architecture.

The *sx?-nec-superux* target is compatible with the native SX toolchain: the object files produced by the GNU assembler can be used with the SX linker, and on the other the native object files and libraries (such as the standard C library) can be linked with the GNU linker.

3.1.2 GCC

Porting of GCC was based on gcc 4.2.2 and required producing the backend that optimizes the RTL intermediate representation [6] produced by the GCC front-end and finally converts it into an SX assembly source.

RTL is a very low-level intermediate representation (as opposed to preceding GENERIC and GIMPLE intermediate representations), already considering various aspects of the target machine, such as addressing modes, register types and sizes, available instructions etc.

The port required defining various symbols that control the GCC code generation, as well as a machine description file that provides templates for converting RTL into SX assembly. After the initial implementation, the porting effort was test driven—running the extensive GCC testsuite pointed to remaining bugs and flaws, which were in turn fixed, and the process was repeated.

The port showed that it is relatively easy and quick to get to a compiler that produces correct unoptimized code. The RTL optimization passes were hard to get right, though, and debugging and fixing these consumed by far the largest amount of time for the GCC port.

Once again, the port provides two targets, corresponding to the two binutils targets:

- *sx?-nec-superux* target implicitly defines symbols that are expected by Super-UX C headers (*__SUPERUX__ _SIZE_T64_* etc.), and provides Super-UX specific behaviour of *main()* (i.e. running static constructors for C++ on entry, calling initialization code from Super-UX standard C runtime etc.). It also respects the Super-UX calling convention, making it possible to mix code built with GCC or with the native compiler.
- *sx?-nec-linux* target provides a Linux-like (better: glibc-like) behaviour of *main()*. Currently, the calling convention is the same as for the Super-UX target, this is, however, expected to change in the future.

3.1.3 Current Toolchain Status

The binutils tools are fully ported and can be used to operate on the libraries and object files produced by either the native compiler or SX port of GCC.

The SX port of binutils is currently being reviewed and discussed with the upstream binutils maintainers in order to push it to the mainline binutils source tree.

Table 1 shows testsuite results for three main ported functionalities of the compiler suite: gcc (the C compiler), g++ (the C++ compiler) and libstdc++v3 (the standard C++ library).

Table 1 GCC testsuite results

	GCC	G++	LIBSTDC++
Expected passes	39319	12089	3330
Expected failures	78	65	13
Unexpected failures	4	105	254
Unexpected successes	2	0	0

The C compiler can be considered functional, and can and is used to build both the Kitten kernel for SX machines and user space software for Super-UX.

The C++ compiler can build functional basic C++ code. Complex C++ constructs combined with heavy optimization can still cause bugs in the generated code. Currently work to improve the C++ compiler to produce proper code for difficult construct is ongoing.

The compiler does lack some functionality that a production quality compiler needs, but that for our purpose, building a kernel can be ignored :

- debugging information is incomplete; furthermore, GCC does not support C++ debugging information to be provided in the SDB debugging format that the Super-UX debugger understands. Implementing better support for the (obsolete) SDB debugging format is not worth the effort required: implementing DWARF-2 or at least STABS debugging formats, and porting the GNU Debugger (GDB) to Super-UX is a better choice.

- profiling is not supported.
- thread-local storage (TLS) is not supported.
- generation of position-independent code (PIC) is not supported.
- instruction rescheduling is not supported: code performance could probably be somewhat improved by implementing it.
- vector instructions are not supported at all. GCC does, however, provide some basic auto-vectorization support that could be leveraged to produce vectorized code from C sources.

All these shortcomings can be fixed with an appropriate effort invested in further improving of the compiler.

Ada, Objective C and Fortran compiler front-ends have so far not been ported.

3.1.4 Future Work

Future work on the toolchain includes (but is not limited to):

- finalizing the C++ port,
- forward porting GCC port to the current GCC sources, and pushing the SX port upstream to the GCC mainline source tree,
- implementing an ELF toolchain (to be used with an ELF based kernel for SX),
- porting the missing functionality in GCC (see Sect. 3.1.3),
- implement a register calling convention (beneficial for building kernel, where performance of scalar code is of utmost importance).

3.2 User Space and I/O Forwarding

Providing a free user space environment on top of Super-UX is important for enabling the support of a wider variety of applications and simplifying testing. For example applications that are developed and built on Linux laptops are expecting functionality that can be found in a Linux and glibc environment. New scalable filesystems are becoming important and models where external I/O servers are used to offload compute nodes are becoming more common.

Operating systems have a level of standardization under POSIX, where system calls and arguments are defined. Still there are differences between Super-UX and Linux that need work to overcome.

3.2.1 Newlib

As gcc and binutils became stable the need for a free C library became important for testing. Initial porting of newlib [7] for Super-UX was made in the fall of 2008, to remedy the immediate need of a free C library, and providing a possibility to have more control over calls like *fopen* and *fprintf*.

Newlib is targeted for embedded systems, which makes it easier to port than for example the larger glibc. It is limited to basic libc functionality and does only provide the following headers: stdlib.h, wchar.h, stdio.h, signal.h, ctype.h, time.h, string.h, locale.h Newlib also provides a mathematical library (libm.a and math.h). Sockets are one example of missing functionality.

Newlib builds on libgloss, which is also a part of the newlib distribution. Libgloss talks to the operating system by invoking the actual system calls. The work started with comparing the arguments of the limited number of system calls needed for libgloss. The arguments are handed into the Super-UX kernel by setting up the correct arguments and calling the *monc* assembler instruction causing a trap (interrupt) in the kernel. For the few system calls needed the argument format for Super-UX was consistent with that expected by newlib. Some input values (e.g. flags for the open call) and output values (returned error constants) had to be adapted in order to match Super-UX.

With the initial Kitten kernel port, a user space C library was also needed, to provide basic I/O. As much of the work already had been done adding a SX Kitten variant was easy. The few system calls that Kitten provides were enough to have a small functional C library. Figure 2 sketches the layers below an application using newlib: the green diamonds are the library calls invoked by the application, which are either handled purely in library functionality (e.g. for mathematical functions) or translate in libgloss to system calls (the dark blue diamonds) into the Super-UX kernel.

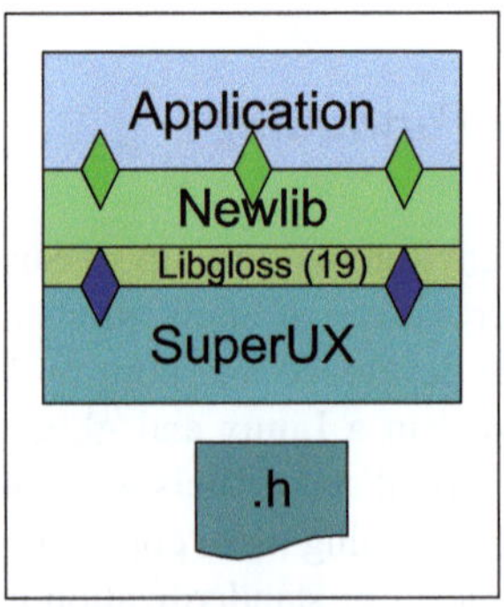

Fig. 2 Component layers for an application linked with newlib on top of Super-UX

Future of Newlib

The limited scope of newlib is a good start for testing, but eventually glibc will be needed to support larger application needs. A future step is also to get the newlib port to a state where it can be released, used by users and pushed to mainline newlib.

3.2.2 Virtualization Layer

To provide the users of NEC SX vector computers with a Linux view, work on a virtualization layer for glibc is being investigated. The idea (Fig. 3) is based on providing a x86_64 Linux-flavored glibc with the related include files, calling functions in a syscall virtualization layer instead of Super-UX system calls. Some of these Linux-like system calls will translate directly to Super-UX calls, some will need intermediate code to be adapted, and some will not be implementable. The last class will simply return "no existing system call" (ENOSYS).

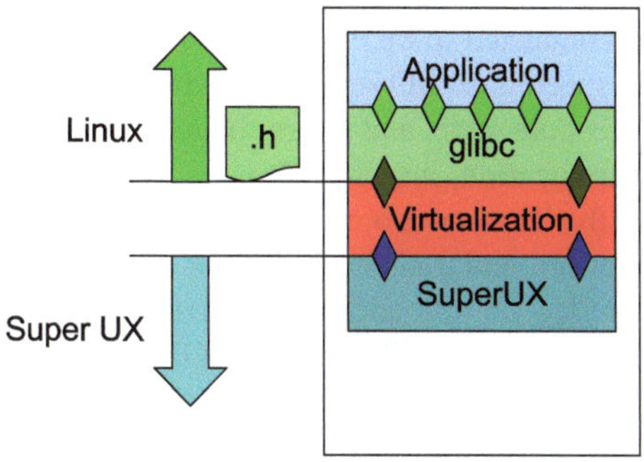

Fig. 3 Glibc x86_64 on top of Super-UX

After initial investigation three different classes of problems were found, as sketched in Fig. 4: The first class contains calls where the arguments do not have to be changed in a typing respect. Looking into the details the content still might have to be changed, like the values of input flag constants or output error code constants. The second class refers to calls where no one-to-one corresponding all exists in Super-UX and Linux, but that could either be implemented by using several Super-UX system calls or by emulating the call in the virtualization layer. The last option in this calls is to fail and return a ENOSYS error. The last class is not really relevant for the virtualization layer, but needs consideration. The system calls that Super-UX provides but that do not have any similar call under Linux.

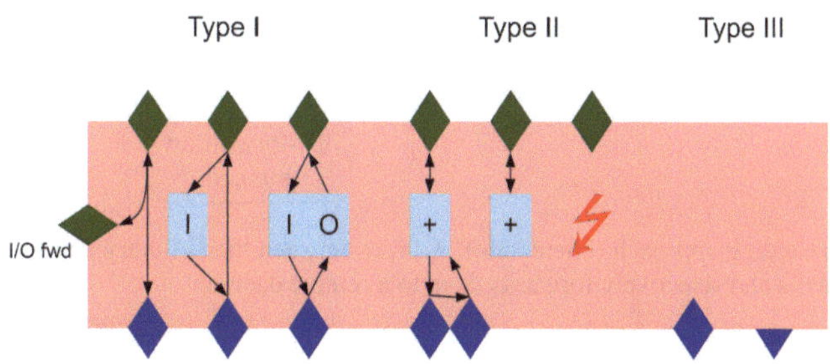

Fig. 4 Different classes of problems when building a virtualized glibc

All of the Linux system calls that make sense could also be forwarded to Linux nodes where they could be executed in an Linux environment. Primarily system calls for I/O come to mind to provide access to Linux fileservers.

3.2.3 I/O Forwarding

With growing sizes of clusters the concept of I/O forwarding has become more attractive because it allows a specialization of cluster nodes e.g. into compute and I/O nodes. Specialized nodes can do their work more effectively and lead to less expensive solutions, for example when compute nodes are diskless or when only I/O nodes have to be connected to an expensive storage area network (SAN). The general idea is to let a specialized I/O server node execute the file I/O operations locally, on behalf of a compute node, and return the results to that node. This happens transparently for the application, that doesn't need to even notice that files are not really accessible directly. The core component to I/O forwarding is a fast network to invoke the I/O operations and transfer arguments and data buffers across nodes.

Examples of forwarded or offloaded I/O are: the IBM BlueGene CIOD, the CRAY XT3–XT5 systems using DVS, and the SONY Playstation 3 doing system call forwarding for the SPUs to the PPU processors. There are several ways to implement mechanisms for forwarding system calls or I/O operations, this can be done in the kernel, below the VFS layer (CRAY's DVS), in user-space, through a dedicated network (BlueGene) or through DMA (integrated into newlib on the Cell CPU). We have decided for a user-space approach integrated into a C library that redirects system calls to a daemon running on I/O server nodes. The client side is sketched in Fig. 5.

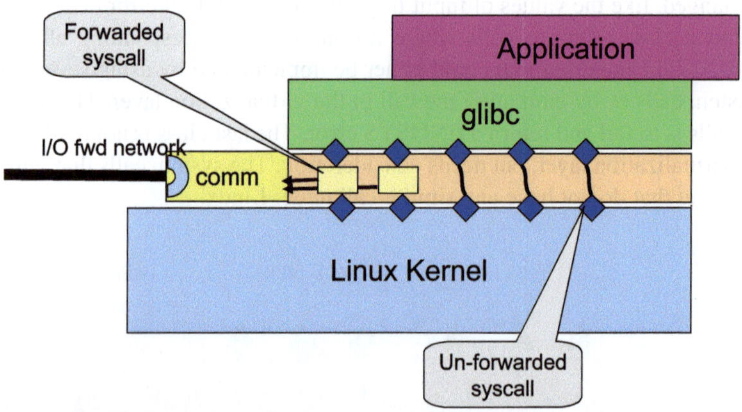

Fig. 5 I/O forwarding approach (client side). A layer between the C library and the kernel intercepts system calls and selectively forwards them to a remote daemon

I/O Forwarding—Current Implementation

The current design provides access to a filesystem with the help of I/O server nodes. These I/O server nodes have a running filesystem client like Lustre and provide access to it over the network for the NEC SX-9 machines. Therefore a Remote Procedure Call (RPC) API has been designed and developed to send and receive messages between the I/O nodes and the compute nodes. It contains a few functions to provide all necessary features and is more or less transparent and independent from the underlying network which can be Gigabit Ethernet, Infiniband or whatever. It can be easily extended with other network devices and network protocols. The RPC API also handles different endianness of SX and the I/O server node. The implementation is designed that every user process on the SX-9 has it's dedicated I/O server (iosrv) daemon on the I/O server node.

The RPC API is called iofwd and is implemented as a user library. Beside of it there are more components which are needed for the application to use I/O forwarding. Figure 6 gives an overview.

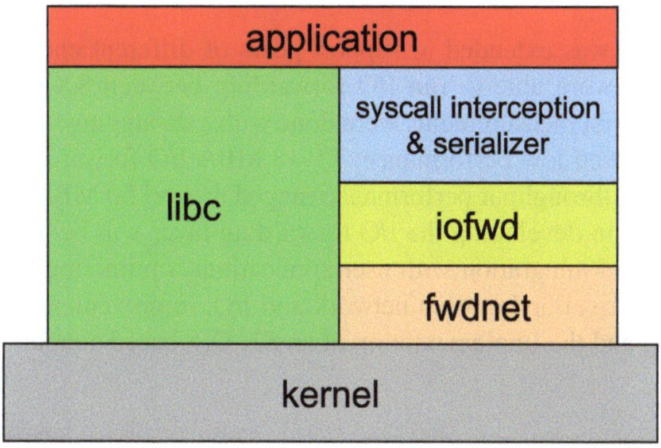

Fig. 6 I/O forwarding client architecture

The underlying communication interface is implemented in the fwdnet library and currently based on GASNet, Global-Address Space Networking [8]. GASNet supports many architectures and different network devices, also called conduits. It can be exchanged with other interfaces like Sandia Portals [9]. Above iofwd there is a library which intercepts the desired I/O system calls and serializes them on the compute node for sending (forwarding) and deserialize them on the I/O server node after receiving the call. Then the I/O server executes the actual system call and returns the result and the standard output/error back to the compute node.

I/O Forwarding Library Status

The GASNet communication interface is ported to the SX-9. The iofwd library is fully implemented and tested and shows that the approach basically works. The serializer and intercepter originally code is inspired from the ZeptoOS ZOID project that aimes at replacing the IBM compute node environment on the Blue Gene.

First I/O forwarding was tested between two Linux machines successfully, the results measured with 8 MB I/O size are summarized in Table 2. Most I/O system calls were tested and an *iozone* run was performed over the I/O forwarding link, confirming that the concept works correctly, though it's performance is improvable.

Table 2 Linux-to-Linux I/O forwarding, initial results

	I/O fwd [MB/s]		bare GASnet [MB/s]	
	read	write	receive	send
Gbit-ethernet	58	49	113	113
IP over IB	143	107	275	491

After GASNet was extended to support peers of different endianness with the UDP conduit we were able to run I/O forwarding between SX-9 and Linux I/O server node. The first measurements were done with a debugging version of GASnet and therefore showed low performance: 17–18 MB/s I/O forwarding performance, while the GASnet throughput performance ranged around 50 MB/s.

The next steps in developing the I/O forwarding layer will be: the improvement of the usability and integration with user applications, optimizing the performance by improving the overlap between network and I/O, improvement of the error and failure handling and the implementation of an SX IXS based conduit.

3.3 Kernel

As explained in Sect. 2 the direct porting of a Linux kernel was considered too risky at an early stage of the project. Alternatives were considered: microkernels like L4 [10] or light weight kernels. L4 requires a C++ compiler that supports inline assembly like g++ from the GNU compiler collection, which was actually considered minor and a side project in the SX GCC development path. A light weight kernel like Kitten [2] requires only a gcc compiler producing correct code. Therefore right after the SX GCC port started to produce code of decent quality we started porting the Linux based Kitten light weight kernel. This was planned as a proof of concept for core kernel components that we had to write for the first time on an SX CPU: bootstrapping a CPU, interrupt handling, atomic operations and spinlocks, memory management, context switching and system calls implementation.

3.3.1 Kitten LWK

Kitten is a very simple kernel derived partly from Linux code and implemented for x86_64 at Sandia National Labs. It is not intended to be a general purpose kernel and provides only very basic resource management functionality like

- process management: process creation, per CPU run-queue, but no cleanup of process memory at exit,
- memory management: tracks contiguous regions of physical memory, a part is dedicated to the kernel while the rest can be used by user space and managed from user space,
- interrupt handling: no timer ticks,
- few devices but no generic PCI infrastructure, no block devices,
- light weight TCP/IP stack.

POSIX filesystems are not implemented in the kernel and need to be handled from user space, either through library implementations of the filesystem client code or through I/O forwarding.

Process memory allocation, process loading, scheduling and cleanup are handled from user space as well, very much similar to the servers concept of microkernels. In order to keep the OS behavior deterministic and the OS interruptions (jitter) short and predictable the user space part of the resource management is concentrated in one single thread, for example in the process control thread in the Catamount LWK [11].

3.3.2 Implementation and Status

Bootstrapping

In porting Kitten to the SX architecture, a very much different architecture than the x86_64 architecture, the design and implementation had to be carefully crafted. The first major hurdle of at all starting the porting was to find the proper format and place for an executable kernel, that would run on the SX at boot. By completing the SX COFF format in binutils and writing a sfs filesystem module for dealing with SX partitions, and MFF ramdisks an executable could be loaded and started from the SX initial system loader (ISL).

Early Introspection

For inspecting early kernel activity we faced the problem of communicating with the kernel and missing an equivalent of an early console and *kprintf()* call. This was first solved with the ISL ability to dump memory to disk and then with a special program read the memory dumps. By providing a special buffer in a known place in memory, the printouts could be read after a memory dump. Basic printing capability was then added by communicating with the ISL in console mode, sending the text to the external terminal connected to the SX. This simplified the debugging, but the memory buffer method was still needed when dealing with interrupts. Communicating with the external console is not possible in interrupt mode of the kernel, as the communication itself needs interrupts to function.

Stack and Memory Layout

In mapping the SX memory hardware into Linux concepts some considerations had to be made on how the memory and stack layout would be handled. The stack layout had to stay as compatible to the Super-UX code as possible, in order to be able to mix code produced by the SX GCC compiler with that of the highly optimizing sxcc and sxf90 compilers. The Super-UX calling conventions pass arguments to function calls on stack, we kept this convention in order to facilitate also the passing of arguments into the kernel in system calls. The stack layout is depicted in Fig. 7. Register \$s0 points to the top of the stack, \$s1 is the frame pointer of the current function and points to the top of the RSA (register storage area) of the caller function. The register \$s2 is the stack pointer, and leaves space for the next (callee) RSA.

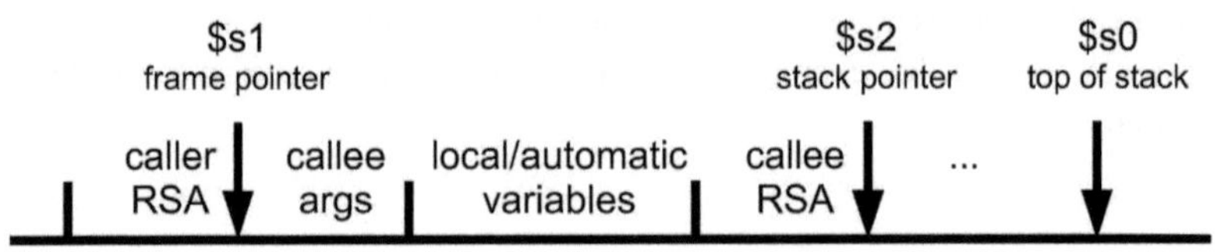

Fig. 7 Stack layout in Kitten LWK

The memory management on the SX vector CPU is quite different from that of normal scalar CPUs. Vector instructions are SIMD operations involving up to 512 64 bit entities in one instruction, and recovering from faults occurring in the middle of vector operations is not possible in an effective way. Therefore in vector mode page and TLB faults are unrecoverable and should not occur. The virtual memory management of the SX CPU is optimized for this behavior (sometimes called "real memory mode") and is based on partial space tables (PSPTs) that represent a contiguous piece of virtual address space and contain 256 page table entries (PTEs), each representing a physical page. While the SX9 CPU can manage pages of different sizes (32 KB, 1 MB, 4 MB, 16 MB, 64 MB and 256 MB) all PTEs in a PSPT point to pages of same size. A SX CPU can load into its address translation buffer (ATB) 40 PSPTs for managing local memory, and a few more for doing address translation of remote node memory. The virtual address spaces represented by PSPTs can not overlap. Physical and virtual address space is 40 bits wide on SX6-SX8, this has been increased on SX9 where physical memory can be represented by 42 bits and virtual address space is as large as 256 TB (48 bits).

The limitations of the memory management hardware compared to general purpose scalar processors map well onto the simplified memory and address space management of the Kitten LWK. This fully loads an address space of a process when switching context in order to avoid page and TLB misses. We implemented a three level pagetables structure with the lowest level directory being exactly the PSPT with 256 PTEs. For SX6–SX8 we have defined a virtual memory layout which maps the first 1 GB of physical memory identically at address 0x0, and maps the first 16 GB of a machine linearly at the address 0xf800000000. The ISL code is mapped at 0xf800100000 and the kernel code is loaded at 0xf800900000. User space programs are loaded at the address 0x400000000. In the initial implementation of Kitten for the SX we decided to design and implement normal and huge page tables using 4 MB pages as normal pages and 16 or 64 MB pages as huge pages.

Interrupts

Interrupts on the SX architecture gets signaled via a 64 bit interrupt vector. The NPSB (next process status block) register in the CPU points to the context in which the interrupt is handled, the first 32 scalar registers of the new context are loaded automatically in hardware, while storing the current context in a piece of memory to which the CPSB (current process status block) register points. We decided to use two kinds of interrupt contexts: one for CPUs running in kernel mode, a common context to any kernel threads, and a user-space thread specific interrupt context attached to each processes task descriptor structure.

System Calls

System calls are implemented by triggering a *monc* assembler instruction in the user space library. The monc instruction raises a trap in the CPU and causes the CPU to switch to kernel context. The arguments of the latest subroutine call together with the argument of the monc instruction are available to the kernel, and with that information the proper action can be performed by the kernel.

Context Switch

A context switch is a rather expensive operation because it needs to save all 128 scalar registers, a significant number of special registers (e.g. performance counters) and the content of the vector and vector data registers. This easily adds up to 256 KB of context to be stored/loaded. For optimizing context switches in a LWK we decided to try reducing the amount of context to be saved:

- The kernel code was limited to use only scalar registers \$s0 to \$s34 and \$s123 to \$s127. This is easily doable with the own SX GCC compiler.

- In the current implementation the kernel code does not use vector registers. We need to evaluate the impact of this decision on performance, and still have the alternative to store vector register context lazily in future if it turns out to be needed for optimized *memcpy* and similar routines.
- Use the return from interrupt automatism for switching context without invoking an additional interrupt.

With these optimizations switching from user space to kernel context and back is much cheaper than under Super-UX. Running the SX machine as a compute node with one thread dedicated to each CPU and almost no other OS related user space processes also avoids switching between different user space contexts, thus saving the time needed for vector context load/store.

User Space

The final step in the kernel startup is passing over control to the very first user space process usually called *init*. The x86_64 implementation of the Kitten LWK receives the initial ramdisk (initrd) from the boot loader in a special area of memory. This piece of memory is interpreted by Kitten as an ELF executable, loaded and started in user space as the first user space process. Since we don't have control over the SX ISL code, we can not modify it to pass in an *initrd* at boot time. The workaround for this limitation was to embed a COFF user space executable as a special binary object into the data section of the Kitten kernel at link time, and load it into the appropriate place in memory at the end of the kernel startup.

Status

Thanks to the simplified testing platform SX-6i, tests could be started early. The SX-6i is a one CPU SX machine where the scalar instruction set is equal to the SX-8. The one CPU platform was also ideal to start with for the simple port as SMP issues could be disregarded for the first steps.

With this platform, steady progress was made and Kitten now runs up to starting an user space application. To be able to support this, memory management was implemented, context switch between several threads and timers are working. The interrupt handler treats the most common interrupts and that subset of interrupt is enough to run a kernel. The LWIP stack has been enabled and the LWIP thread is working. System calls handling have been enabled, and implemented, for all system calls supported by Kitten. For example a `printf` to standard out from a user application gets forwarded through the kernel to the external console. Via the newlib port, the Kitten system call layer and at the end communicated to the console with the help of the interrupt handler.

For the SX-6i work with the PCI bus and its card were also started but discontinued. Although progress was made towards a working network card, the work was put aside in preference for switching testing to the SX-8. SX-8 has a different PCI structure, hence further work on SX-6i would not have been beneficial. Still parts of the work can also be used for the SX-8.

Initial tests on the SX-8 proved to be surprisingly successful and the kernel worked up to the point of starting and running the user application.

4 Outlook

The Kitten LWK turned out to be a very promising path towards a compute node kernel for the SX, therefore we will improve its capabilities and for the near future not turn towards porting a basic Linux kernel. With the successful testing of the code base on the SX-8 the next step is to turn on SMP. Most part of the SMP support has been added but never tested. Booting secondary CPUs needs to be investigated and implemented. After that a network device driver for the SX8/SX9 needs to be implemented, this could be done in conjunction with work on adding a generic PCI infrastructure. A collaboration with Sandia National Labs on further development of the Kitten LWK is desirable and expected to start soon.

In order to use the SX Kitten LWK we started design and implementation work on a user space process control thread that will manage the resources from user space, receive requests to load/start/kill user space programs from control nodes, and replace most of the user space usually available on a full blown OS. This development is currently being done on x86_64 hardware and will work as a simplified user space also on normal Linux clusters requiring low OS noise.

On the user space library side of the project current efforts focus on producing a usable production level infrastructure for SX users that allows them to access filesystems mounted on Linux servers directly from programs running on the SX9 under Super-UX. In a later step this effort will be complemented by providing a Linux-like glibc with a virtualization layer including I/O forwarding. This will ease compilation of programs developed under Linux and provide a path forward for moving towards light weight kernels.

The current developments open also some interesting paths to side projects:

- Researching hybrid programming paradigms that become much easier to implement once the I/O forwarding infrastructure is working stable.
- Implementing PGAS languages like UPC or Co-Array Fortran on top of the network infrastructure used for I/O forwarding.
- Work on a vectorizing GCC, join similar efforts done for other architectures.

Concluding, after a year of work the project has lead to very concrete results: working binutils and GCC ports to the SX architecture, the development of an I/O forwarding library on top of the GASnet network layer, a simple free libc implementation and a working light weight kernel on the SX. This has been achieved despite rather limited access to in-depth documentation and with a team of 6 people with various backgrounds. The embedding into the Teraflops Workbench project has exposed the team to interesting real-life problems and extended the scope of the developments. The developments couldn't have been this advanced after the short time without access and use of free open source software, a programming paradigm that we want to continue to follow.

References

1. Roller, S.P., Multi-scale and multi-physics applications user requirements for future applications. High Performance Computing on Vector Systems, pp. 45–52 (2008). doi:10.1007/978-3-540-85869-0_5. http://www.springerlink.com/content/q23v073177672208/
2. The Kitten Lightweight Kernel. https://software.sandia.gov/trac/kitten
3. The Linux Kernel Archives. http://kernel.org/
4. Hagen, W. von: The Definitive Guide to GCC. Apress (2006), ISBN 1590595858
5. GNU Binutils. http://www.gnu.org/software/binutils/
6. GNU Compiler Collection (GCC) Internals. http://gcc.gnu.org/onlinedocs/gccint/
7. The Newlib Homepage. http://sourceware.org/newlib/
8. GASNet. http://gasnet.cs.berkeley.edu
9. Sandia Portals. http://www.cs.sandia.gov/Portals/
10. Liedtke, J., Toward real microkernels. Published in Communication of the ACM (CACM), vol. 39/9, pp. 70–77, September 1996
11. Kelly, S. , Brightwell, R., Software architecture of the lightweight Kernel, Catamount. In: Proceedings of the 2005 Cray Users' Group Annual Technical Conference, Albuquerque, New Mexico, May 2005.

Development of APIs for Desktop Supercomputing

Guehee Kim, Yoshio Suzuki, Akemi Nishida, Hiroshi Takemiya

Abstract We developed the Script Generator API to support Grid users to develop their own Grid-enabled client application. The Script Generator API automatically generates a Grid-enabled workflow script needed to execute jobs on a Grid system. The Script Generator API enables users to develop their application with the complex job flow which cannot be generated using existing workflow tools. We implemented the Script Generator API in our Grid infrastructure and utilized it to the three-dimensional virtual plant vibration simulator. By developing Grid-enabled client application for the three-dimensional virtual plant vibration simulator, we confirmed the usability of the Script Generator API.

1 Introduction

With the spread of the Grid computing [1], more and more scientists and engineers pay attention to the Grid computing and hope to solve their problems and carry out their R&Ds using not only computational resources but also data, storages, networks and experimental sensors. Up to now, various types of Grid application have been developed and can be classified into distributed computing application, data intensive application, parameter survey application, remote collaboration and so on by their computation and data processing patterns. At present, almost of these applications are constructed on special-purposed Grid portals through collaborative works between computational researchers, IT engineers and Grid technicians. On the other hand, some researchers want to have their own Grid-enabled applications on not the Grid portal but their desktops, as if they have desktop supercomputers (hereafter, we call this type of Grid user and developer environment as GDS (Grid-enabled Desk-

Guehee Kim, Yoshio Suzuki, Akemi Nishida, Hiroshi Takemiya
Center for Computational Science and e-Systems, Japan Atomic Energy Agency, 6-9-3 Higashi-Ueno, Taito-ku, Tokyo 110-0015, Japan, e-mail: kim.guehee@jaea.go.jp, suzuki.yoshio@jaea.go.jp, nishida.akemi@jaea.go.jp, takemiya.hiroshi@jaea.go.jp

M. Resch et al. (eds.), *High Performance Computing on Vector Systems 2009*, 97
DOI 10.1007/978-3-642-03913-3_9 © Springer-Verlag Berlin Heidelberg 2010

top Supercomputing)). Since those researchers are not always Grid technicians, it is important issue that GDS provides robust tools with which uses can easily make the fullset use of Grid services. The tool which enables users to develop their applications on GDS is useful.

In the early days of R&D of Grid computing, the Globus toolkit 2 was de facto tools to construct a Grid environment. Grid sites such as GriPhyN [2], FusionGrid [3], and PPDG [4] provide primitive tools such as toolkits, APIs (Application Program Interfaces) and libraries for Grid users and application developers. However, those support tools were so low level that Grid users and application developers had difficulty with using those tools. Furthermore, users must be careful to manage their Grid membership information such as authority certificates or grid-mapfiles for Grid authentication. Later, to provide a standardized Grid technology for Gird users and application developers, the concept of OGSA (Open Grid Service Architecture) has been proposed by GGF (Global Grid Forum) [5]. As implementations of OGSA, the Globus toolkit 3 and UNICORE 6 [6] appeared and enabled developers to construct Grid systems with commodity web technologies. Grid portals such as NAREGI [7], NEESgrid [8], ESG [9], and JVO [10] are the Grid sites constructed on the basis of the Globus toolkit 3. At present, some Grid portals provide higher-level Grid application development tools that considerably hide the complexity of the Grid infrastructure. Also, user environment using web browser allows users to easily access the Grid system without a troublesome management of authentication information.

Although all of these progresses of Grid computing, existing Grid systems do not provide sufficient tools for GDS. Use of the tools is still hard burden for end-users. Grid-enabled workflow tools developed so far are useful to generate applications which have strong dependency between jobs and data remotely distributed. But those tools have limits to develop applications with a complex job flow. To break through this status, we have developed a Grid-enabled client APIs for researchers who want realize GDS. These client APIs have not only low level functions which are responsible for providing basic Grid services but also high level functions to support development of GDS applications which are composed of complex jobs and data controls.

This paper is composed of the following sections. In Sect. 2, we show the client APIs developed so far and explain the Script Generator API in detail. The most distinguished feature of the Script Generator API is automatic generation of a Grid-enabled workflow script. We implemented the client APIs in our Grid infrastructure, AEGIS (Atomic Energy Grid Infrastructure). In Sect. 3, we describe the development of GDS application for "three-dimensional virtual plant vibration simulator" using the client APIs. In Sect. 4, we summarize our work.

2 Client APIs for GDS

2.1 Client APIs

So far, we have developed client APIs, which are divided into three groups by their service level as shown in Table 1. At the low level, Communication API and Authentication API are located. These APIs establish an authenticated and encrypted communication route from a GDS terminal to a Grid system. At the middle level, File Control API, Job Control API, Resource Information API and Script Generator API supplement connections between low level APIs and high level APIs. Script Generator API will be discussed in detail in Sect. 2.2. Fault Tolerance API and Parameter Survey API at the high level involve the substitution of computing resources for fault tolerance and dynamic resources allocation for parameter survey, respectively. Since the client APIs are provided as a software package in C language, it is simple and easy to install them into a user desktop. By compiling and linking these APIs with user programs, users can easily develop GDS applications.

Table 1 Client APIs for GDS

Level	Client APIs
High	Fault Tolerance API
	Parameter Survey AI
Middle	Script Generator API
	Resource Information API
	Job Control API
	File Control API
Low	Authentication API
	Communication API

2.2 Script Generator API

Script Generator API offers four services; user interface, staging, script generating, job status monitoring. The most important service is the script generation service which automatically generates Grid-enabled workflow script. Because of this service, GDS users need not use Grid-enabled workflow tools, which realizes more user-friendly GDS environment and enables GDS users to develop more complex applications without the limitation of workflow tools. Figure 1 shows the schematic diagram of the Script Generator API. Here, Basic APIs are including five client APIs: Communication API, Authentication API, File Control API, Job Control API, and Resource Information API. To submit jobs from GDS application, users assign job attributes such as host names, batch queue names, program names, parameters and so on. Next, users demand job execution from the GDS application. The demand is transferred to the Script Generator API which generates a script and handovers it to the Basic APIs. Finally, the Basic APIs execute the job in liaison with a Grid system. Figure 2 shows behaviours of the four services of the Script Generator API between GDS application and the Basic APIs.

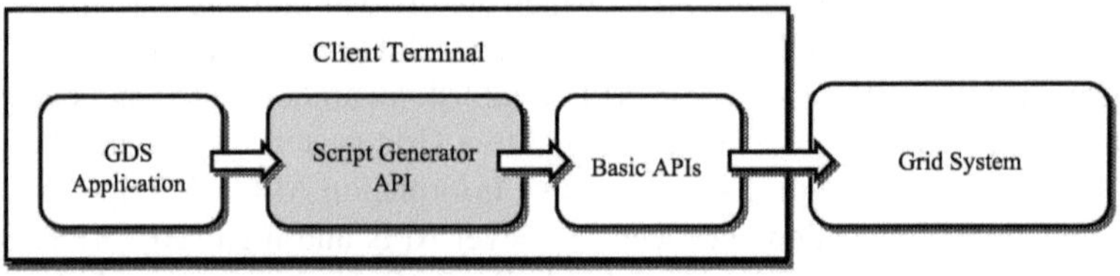

Fig. 1 Schematic diagram of the Script Generator API

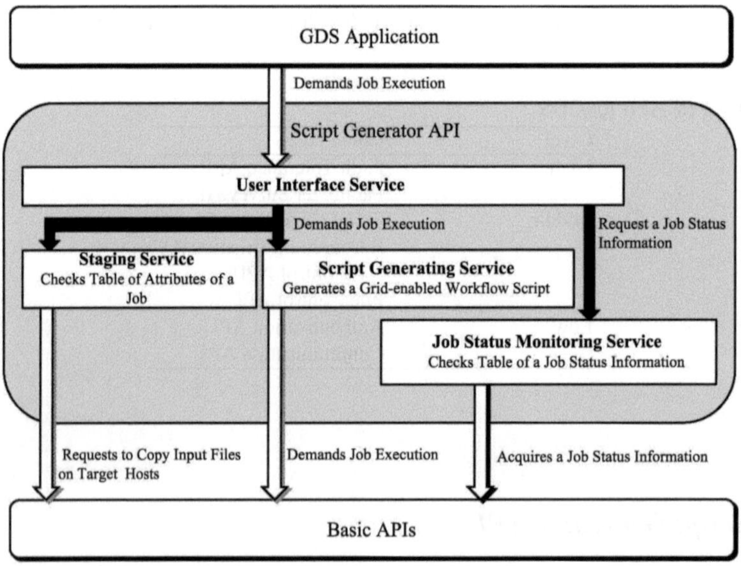

Fig. 2 Schematic diagram of four services of Script Generator API: User Interface service, Staging service, Script Generating service, and Monitoring Job Status service

- *User Interface Service* The user interface service invokes three other services of staging, script generating, and job status monitoring. It has 13 functions as follows:

  1. **CreatJob** This function generates an object of a job.
  2. **SetJobAttribute** This function sets a table of attributes of a job.

3. **EntryJob** This function groups multiple jobs and registers them as a "multijob", an execution unit. One job can be registered as a multijob. Using this function, it is possible to design complex application.
4. **CreateMultiJob** This function generates an object of a multijob registered by the EntryJob function.
5. **SubmitJob** This function requests execution of a multijob to the Basic APIs.
6. **WaitForAnyTerminated** This function waits for any job of a multijob to be terminated.
7. **WaitForAllTerminated** This function waits for all jobs of a multijob to be terminated.
8. **GetMultiJob** This function returns a multijob object.
9. **CancelJob** This function cancels execution of a multijob.
10. **DestroyMultiJob** This function deletes the object of a multijob.
11. **GetJobInfo** This function provides job status information of a submitted job with the information such as a target host, start time, elapsed time and so on.
12. **DestroyJobInfo** This function deletes job status information of a job acquired by the GetJobInfo function.
13. **DestroyJob** This function deletes status information of a job when it is canceled or terminated.

- *Staging Service* This service checks whether input files exist on target host and requests the Basic APIs to copy the input files of the user‘s desktop on the target host, which is inevitable procedure for job execution and is very usable when a number of input files must be prepared.
- *Script Generating Service* After a GDS application demands job execution, this service generates a Grid-enabled workflow script and requests the Basic APIs to execute jobs. These scripts include information of attributes of jobs and how the jobs are related with each other. Each Grid system has its own unique script format and provides its own dedicated workflow tool. Problem is that use of workflow tools is time-consuming and diversity of complex job control is generally impossible. The script generating service allows users to execute their jobs without workflow tools, which is very usable for a large-scale simulation with complex job flow.
- *Job Status Monitoring Service* This service regularly gets job status information and provides it with GDS application. A user can monitor job status and restart immediately when any job is terminated abnormally.

2.3 Implementation of Script Generator API in AEGIS

CCSE/JAEA (Center for Computational Science and e-Systems of Japan Atomic Energy Agency) launched the R&D of AEGIS to develop computational infrastructure for atomic energy research [11, 12]. The R&D aims to establish predictable science and engineering to guarantee security and safety of atomic energy and to preserve the environment by constructing a real-scale nuclear facility. To develop the base of AEGIS, a network computing system is constructed by expanding the knowledge and skills of ITBL (Information Technology Based Laboratory) infrastructure software [13, 14].

We implemented the Script Generator API and the Basic APIs in AEGIS as shown in Fig. 3, where arrows with numbers mean processes by services of the Script Generator API. Processes from 1-1) to 1-5) are needed for job execution, where user interface service, staging service and script generating service are invoked. Processes from 2-1) to 2-3) are for monitoring job status, where user interface service and job status monitoring service are involved.

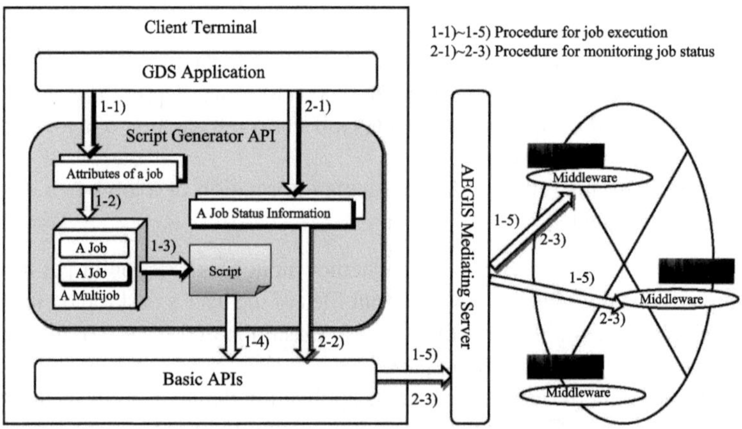

Fig. 3 Schematic view of implementation of Script Generator API and Basic APIs in AEGIS

Each process is explained as follows: 1-1) A user assigns attributes of a job. When there are multiple jobs to be registered as a multijob, each job attributes assignment is repeated. Then, the user demands job execution. 1-2) Script Generator API registers the multiple jobs (or a job) as a multijob. Simultaneously, The API checks table of attributes of the job(s) to determine whether to prepare input files on target host(s). 1-3) Script Generator API generates a TME (Task Mapping Editor) script of the multijob. Here, TME is the Grid-enabled workflow tool of AEGIS [15]. 1-4) Script Generator API requests Basic APIs to execute the multijob. 1-5) Basic APIs establish authenticated and encrypted communication route to AEGIS system and start job execution according workflow described in the TME script in liaison with the AEGIS. 2-1) A user demands job execution. 2-2) Script Generator API regularly requests Basic APIs to send job status information of a multijob. 2-3) Basic APIs check job status on AEGIS resources and notify the information to the Script Generator API.

3 Development of GDS Application of Three-dimensional Virtual Plant Vibration Simulator

3.1 Three-dimensional Virtual Plant Vibration Simulator

Three-dimensional Virtual Plant Vibration Simulator is a full-scale seismic analysis system to analyze response of a whole digitalized nuclear power plant to earthquake or vibration effect [12, 16]. The system has been developed by CCSE/JAEA. The final goal is to contribute to safe and stable atomic energy supply. Generally, an entire nuclear power plant is composed of 100 thousand to 10 million parts so that the entire simulation is impossible on a single supercomputer due to its hardware limitation. If each part such as a reactor vessel, water coolers, pipes and so on can be treated and processed as independent components with specific function, we can achieve the entire simulation by deploying the components on various Grid computing resources. The Three-dimensional Virtual Plant Vibration Simulator enables such component deployment on the Grid system because it realizes "from parts to the whole" strategy.

Large components such as a reactor vessel are interconnected by smaller components as pipes. The displacement propagation behaviour between the large and small components was investigated theoretically [17]. This study shows that the large components can be treated as a rigid bodies against the interconnecting pipes due to the large difference of sizes. So we model an entire power plant in hierarchal manner, which treats large components as primary level and pipes as secondary level. In this model, the only displacement boundary condition from the primary level to the secondary level is applied. For example, Fig. 4 shows hierarchical model of HTTR (High Temperature Engineering Test Reactor) at O-arai R&D center of JAEA. In this case, reactor, AWAC (Auxiliary Water Air Cooler) and PWAC (Pressurized Water Air Cooler) belong to the primary level and two pipes belong to the secondary level. To effectively deploy the hierarchical model on the Grid, we plan to carry out simulate this hierarchical mode through pipelined processing method

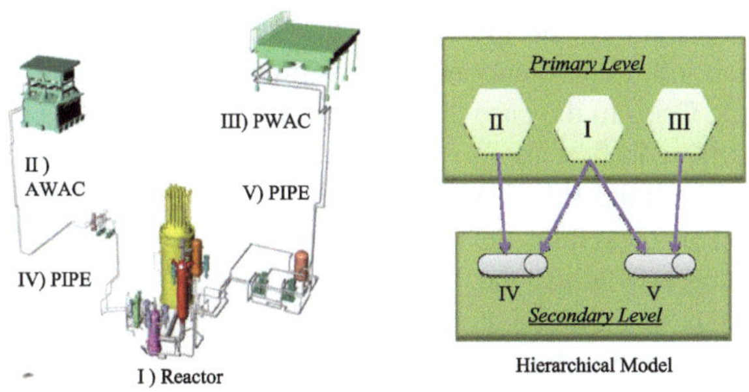

Fig. 4 CAD model of a reactor, cooling systems (AWAC and PWAC) and pipes of HTTR and the corresponding hierarchical model

as shown in Fig. 5. In side of the primary level, simulation advances outputting boundary condition data per each step. Then, the data is immediately transferred to the computers of the secondary level. In side of the secondary level, simulation is progressed by using the boundary condition data as input data per each step.

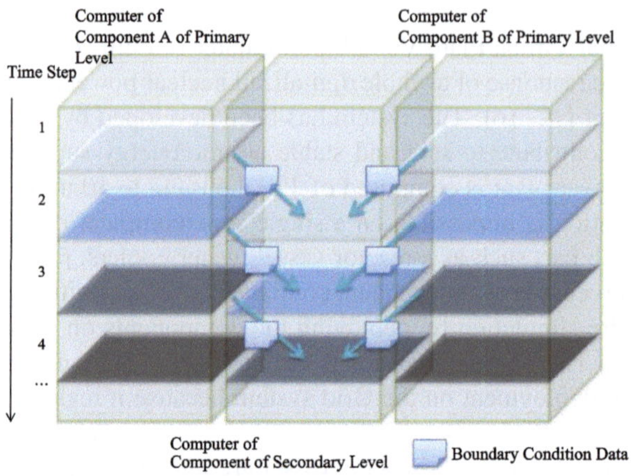

Fig. 5 Pipelined process of the hierarchical model

3.2 Development of GDS Application of Three-dimensional Virtual Plant Vibration Simulator

We developed GDS application of the Three-dimensional Virtual Plant Vibration Simulator using the Script Generator API and the Basic APIs. Users can assign attributes of a job and perform simulations through graphical user interface of the application. Figure 6 shows the simulation of three components of the primary level of the HTTR using the application. In this case, three jobs for the three components are registered as a multijob and submitted to the Script Generator API. According to attributes of each job, three components are simulated using solvers of distributed computers on the Grid. In the case that the Script Generator API is not used, a user must use the TME workflow tool to assign attributes of jobs and define a job flow. Next, the user must import the TME script into a Grid-enabled application. In the case that the Script Generator API is used, the user need not use the TME. For end-users, learning how to use the TME is complicated and time-consuming work. In this respect, the Script Generator API provides AEGIS users with a usable job execution environment. Furthermore, the TME can not define pipelined process. But by using the Script Generator API, we can generate a GDS application realizing that process.

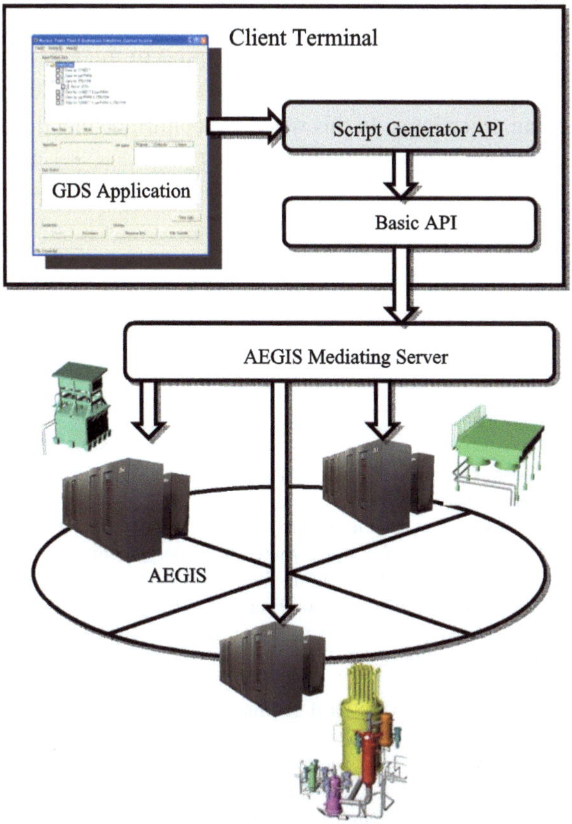

Fig. 6 Distributed and parallel simulation using GDS application

4 Summary

We developed client APIs from low level to high level to support Grid users who want to develop their own Grid-enabled applications on their desktop. The developed client APIs so far are as follows.

1. Low level APIs: Communication API and Authentication API
2. Middle level APIs: Resource Information API, Job Control API, File Control API, and Script Generator API
3. High level APIs: Fault Tolerance API and Parameter Survey API

We implemented them in our Grid infrastructure AEGIS. The Script Generator API automatically generates a Grid-enabled workflow script from a Grid-enabled application. Without the Script Generator API, the workflow script must be manually edited by using Grid-enabled workflow tools. By Script Generator API, users need not use the workflow tools. Furthermore, by the Script Generator API, it is possible to develop applications with complex workflows.

We developed the Grid-enabled application of the Three-dimensional Virtual Plant Vibration Simulator using the Script Generator API. In the previous work, we defined attributes of jobs and job flow using workflow tool. These were complicated and time-consuming procedures. In this work, we confirmed that the Script Generator API which makes these procedures automatic provides a user environment with high usability.

Acknowledgements We acknowledge O-arai R&D center of JAEA for useful discussion and data provision.

References

1. Foster, I., Kesselman, C., et al.: The Grid: Blueprint for a New Computing Infrastructure. Morgan Kaufmann, California (1998)
2. Avery, P., Foster, I.: The GriPhyN project: towards petascale virtual data grids. iVDGL/GriPhyN Technical Reports. http://www.griphyn.org/documents (2000)
3. Schissel, D.P., Burruss, J.R., et al.: Building the U.S. national fusion grid: results from the national fusion collaboratory project. Fusion Eng. Des. **71**, 245–255 (2004)
4. Olson, D.L., Perl, J.: Interfacing interactive data analysis tools with the Grid; the PPDG GS-11 activity. Nucl. Instrum. Methods Phys. Res. Sect. A, Accel. Spectrom. Detect. Assoc. Equip. **502**, 240–422 (2003)
5. Foster, I., Kesselman, C., et al.: The Grid 2: Blueprint for a New Computing Infrastructure. Morgan Kaufmann, California (2004)
6. Riedel, M., Schuller, B., et al.: Web services interfaces and open standards integration into the European UNICORE 6 grid middleware. In: Proceedings of 2007 Middleware for Web Services (MWS 2007) Workshop, pp. 57–60 (2007)
7. Miura, K.: Overview of Japanese science grid project NAREGI. Progr. Inform. **3**, 67–75 (2006)
8. Spencer, Jr., B., Finholt, T.A., et al.: NEESgrid: a distributed collaboratory for advanced earthquake engineering experiment and simulation. In: Proceedings of the Thirteenth World Conference on Earthquake Engineering CDROM:1674 (2004)
9. Foster, I., Alpert, E.: The earth system grid II: turning climate datasets into community resources. In: Annual Meeting of the American Meteorological Society. http://www.earthsystemgrid.org/about/documentsPage.do (2002)
10. Ishihara, Y., Mizumoto, Y.: Construction of Japanese virtual observatory (JVO). FUJITSU Sci. Tech. J. **40**(2), 217–223 (2004)
11. Suzuki, Y., Nakajima, K., et al.: Research and development of fusion grid infrastructure based on atomic energy grid infrastructure (AEGIS). Fusion Eng. Des. **83**, 511–515 (2008)
12. Suzuki, Y., Nakajima, N., et al.: Development of three-dimensional virtual plant vibration simulator on grid computing environment ITBL-IS/AEGIS. In: Proceedings of the Sixteenth International Conference on Nuclear Engineering CDROM:48478 (2007)
13. Higuchi, K., Imamura, T., et al.: Grid computing supporting system on ITBL project. In: Proceedings of the Fifth International Symposium on High Performance Computing (LNCS2858), pp. 245–257 (2003)
14. Fukui, Y., Stubbings, A., et al.: Constructing a virtual laboratory on the Internet: the ITBL project. In: Proceedings of the Fifth International Symposium on High Performance Computing (LNCS2858), pp. 288–297 (2003)
15. Imamura, T., Yamagishi, N., et al.: A visual resource integration environment for distributed applications on the ITBL system. In: Proceedings of the Fifth International Symposium on High Performance Computing (LNCS2858), pp. 258–268 (2003)

16. Tani, M., Nakajima, N., et al.: A methodology of structural analysis for nuclear power plant size of assembly. In: Proceedings of Mathematics & Computation and Supercomputing in Nuclear Applications: the Joint International Topical Meeting CD-ROM (2007)
17. Nishida, A., Liu, P., et al.: Fundamental studies of wave-propagation properties of single layer lattice structures. JSCE J. Struct. Eng. **46B**, 175–179 (2000) (in Japanese)

The Grid Middleware on SX and Its Operation for Nation-Wide Service

Manabu Higashida

Abstract The Cybermedia Center of Osaka University runs a large-scale, high-performance computing system that combines a vector supercomputer with PC clusters. In order that all registered users can equally access this system from a grid environment, the center has been looking at using NAREGI grid middleware. We enhanced the virtual machine feature of the middleware-known as the GridVM (grid virtual machine)-and researched how to effectively supply the resources of the NEC SX Series vector supercomputer to the grid environment.

1 Introduction

The Cybermedia Center of Osaka University runs a large-scale, high-performance computing system that combines a vector supercomputer with PC clusters. In order that all registered users can equally access this system from a grid environment, the center has been looking at using NAREGI grid middleware [1].

As was reported in last year's paper [2] and [3], to adapt NAREGI grid middleware to the system operated at the center, we first upgraded the center's authentication and authorization system so as to create a single sign-on environment that is highly compatible with the existing authentication system, Kerberos (see Fig. 1). We then enhanced the virtual machine feature of the NAREGI grid middleware-known as the *GridVM* (grid virtual machine)-and researched how to effectively supply the resources of the NEC SX Series vector supercomputer to the grid environment. This paper details the results of this research, that is, the development of a program known as the *GridVM for the SX Vector Computer*.

In this paper, we will first introduce the NAREGI grid middleware and describe its features in Sect. 2 and its issues in Sect. 3. We will describe our center's config-

Manabu Higashida
Cybermedia Center, Osaka University, 5-1 Mihogaoka, Ibaraki, Osaka 567-0047, Japan, e-mail: manabu@cmc.osaka-u.ac.jp

M. Resch et al. (eds.), *High Performance Computing on Vector Systems 2009*,

DOI 10.1007/978-3-642-03913-3_10

uration in Sect. 4, and then we will explain how we used the GridVM for the SX Vector Computer to overcome these issues in Sect. 5. Finally we will look at what we have to do to create a more service-oriented grid environment in Sect. 6.

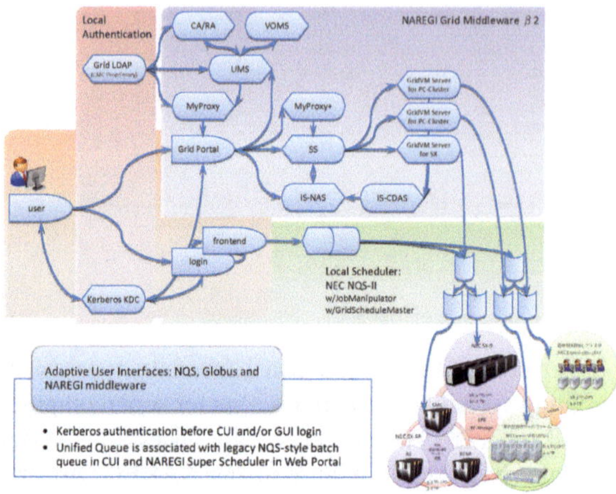

Fig. 1 Deployed NAREGI grid middleware in the Cybermedia Center

2 Structure of NAREGI Grid Middleware

The NAREGI grid middleware manages computing resources by using a combination of three Web services: the Information Service (IS), the Super Scheduler (SS), and the Grid Virtual Machine (GridVM). This section describes the architecture and features of these Web services.

2.1 Managing Resources by Using Web Services: IS, SS, and GridVM

The GridVM virtualizes the computing resources by interfacing with the local scheduler, which actually manages the computing resources, and supplies the computing resources to the grid environment as a Web service. At the same time, resource information defined by DMTF-CIM (Distributed Management Taskforce-Common Information Model) and resource information unique to NAREGI are supplied to the Information Service via the IS-LRPS (Information Service's Local Resource Provider Service). Then, when a job is executed, accumulating information defined by OGSA-RUS (Open Grid Service Architecture?Resource Usage Service) is supplied to the Information Service.

The Super Scheduler functions as a meta-scheduler that aggregates the local schedulers virtualized by the GridVM. The resource information accumulated by the Information Service is searched via a Web service interface defined by OGSA-DAI (Data Access and Integration) and matching information is retrieved and compiled into a GridVM list of job candidates. Job scheduling is then carried out based on the job reservation table of each GridVM node. These job reservation tables are aggregated by the RCS (Reservation Cache Service), which is a sub-module of the Super Scheduler. The most appropriate GridVM node is selected and a job submission slot is reserved. Job description information defined by OGF JSDL (Open Grid Forum?Job Submission Description Language) is then created and a job is specified to be input to the local scheduler via the GRAM (Grid Resource Allocation Manager) Web service interface provided by the GridVM.

The Information Service is a Web service that accumulates and aggregates monitoring and statistical information hierarchically. The Information Service is linked with the Super Scheduler and the GridVM as described above.

2.2 Virtualizing the Computing Resources: GridVM Scheduler and GridVM Engines

The GridVM consists of a GridVM scheduler node, which manages and controls the computing resources via the local scheduler, and a group of GridVM engine nodes, which are the actual computing resources subject to job input by the local scheduler.

The GridVM scheduler's job queue is manipulated as a wrapper of the local scheduler's queue. This wrapper is a plug-in type wrapper, so by replacing the plug-in module, a variety of local schedulers can be used, including PBS Pro, LoadLeveller, NEC's NQS-II, SunGE+PluS/N1GE, Parallelnavi, and LSF. PBS Pro is used here for reference.

The job input to the GridVM engine node group and activated via the wrapper is placed under the management of the GridVM scheduler and status changes are monitored. Until the $\beta 1$ version (pre-released at the end of FY2005), jobs were executed in the sandbox inside the wrapper. However, there was concern that this would lower the job execution performance. Moreover, in view of the fact that NAREGI did not envision highly anonymous users using the system, from the $\beta 2$ version (pre-released at the end of FY2006), the sandbox feature has been made an option. Information from the nodes in the GridVM is collected by the local scheduler and Ganglia and is accumulated in the Information Service via the IS-LRPS Web service.

3 Features and Issues of NAREGI Grid Middleware

The Super Scheduler that executes job scheduling in the NAREGI grid middleware is designed as a 100% reservation-type meta-scheduler. Computing resources are assumed to be managed and controlled by a local scheduler with reservation capabilities, such as PBS Pro or NEC's NQS-II. This enables MPI (Message Passing Interface) jobs that run on multiple GridVM nodes to be executed and improves the work flow.

However, we were requested to change the specifications to enable the execution of non-reserved jobs so that local schedulers without reservation functions could be used, and so that jobs used at existing center operations could coexist with new jobs. But we subsequently found that the scheduling capacity was insufficient because the capabilities of the Information Service were not expanded.

Below are listed the problems we found with the NAREGI grid middleware following extensive study, specifically of the job scheduling features.

3.1 Reservation-Type Job Scheduling

The Super Scheduler executes job scheduling by filling out reservation tables in nodes in the GridVM group extracted by using the information-matching feature. Reservations are made in node units. In large-scale shared memory systems such as the NEC SX or Hitachi SR, where one node uses many CPUs, the scheduling is not necessarily efficient. Also, today's trend toward raising performance by using multi-core systems makes it preferable to allow reservations to be made in CPU core units.

There were also several features that were included in the initially requested specifications but were later deleted. Among these was a reservation table move-up feature, which was essential for raising the operating efficiency of the system. This feature was probably deleted because during development, the Web services messages between the Super Scheduler and the GridVM could not be translated reliably and so the status transitions were not made properly. However, we must find a way to reliably translate messages so as to restore this deleted feature.

3.2 Virtualization and Overheads

The computing resources are virtualized by the GridVM. The GridVM abstracts the processes of the local scheduler, which actually controls the computing resources, enabling creating a grid environment that assumes the existence of heterogeneous systems. In the NAREGI project, because specifications were required that enabled the use of a variety of local schedulers, it was not possible to provide the meta-scheduler with all the features provided by the local schedulers when executing operations such as grasping job status transitions.

The GridVM and the local scheduler communicate via commands specified from the command line and by sending status information via the standard input/output. Some statuses are also transmitted by monitoring the local scheduler's log output.

This has caused operational problems, such as the occurrence of faults due to inconsistencies in the statuses of the GridVM and the local scheduler, and a loss of coherency when either just the GridVM or just the local scheduler is rebooted.

Operations carried out at the center are executed via a vendor MPI, enabling high-performance computing. For this reason, executing jobs in the sandbox is not always desirable. However, the sandbox feature is useful when executing highly anonymous P2P type applications. The center therefore requires the flexibility to select the features that best suit the application at the time.

3.3 Load Concentration on Management Nodes

The Super Scheduler carries a heavy processing load because it handles the WFML and executes information matching via the OGSA-DAI interface. For this reason, there can be multiple Super Schedulers in conjunction with the NAREGI Web portal, dispersing the load. However, even when the Super Scheduler load is dispersed, the reservation map information is managed entirely by the Super Scheduler's RCS sub-module. We considered implementing multiple RCSs, but ended up not doing so before the end of the project.

3.4 Scheduling of Non-reserved Jobs

The Super Scheduler executes 100% reserved job scheduling. The resource provider must therefore supply all the resources so that they belong completely to the NAREGI Super Scheduler. This means that either control of the nodes must be ceded to the NAREGI Super Scheduler, or the Super Scheduler control must be restricted to operations at the center.

The center also needed the NAREGI grid middleware to be able to support local schedulers that did not have reservation capabilities, leading to the addition of a feature to enable the execution of non-reserved jobs. However, when executing non-reserved jobs, the NAREGI Information Service does not receive information concerning factors such as the load average or average queuing time of all the clusters-information that is vital for advanced scheduling. This means that the Super Scheduler can only carry out very simple scheduling (that is, jobs are executed sequentially without referencing a reservation table).

For example, with a PSE (problem solving environment), which is a feature of the NAREGI middleware, the compiling that is required in order to use the application is carried out by the Super Scheduler. However, there is concern that in practice, usability may be lost due to problems such as the lack of a reservation slot, or the input of a job at the bottom of a long queue.

3.5 Maintaining Coherency and Consistency in the Web Services on the Grid

The Super Scheduler and the GridVM are implemented as Web services. All status transitions are carried out by the exchange of SOAP (Simple Object Access Protocol) messages. If a status inconsistency occurs due to a communication problem, the inconsistency must be resolved by executing recovery processing. However, in the NAREGI middleware, several problems remain unresolved due to the difficulty of implementing the required recovery processing. For example, if a reservation cannot be made due to an inconsistency in the reservation maps managed by the Super Scheduler and the GridVM, there should be a way to return the reservation maps to a coherent state and re-execute the reservation. However, the processing required to execute these operations is not included in the NAREGI middleware.

Unresolved issues such as a long wait time have also created operational problems. For example, versions up until $\beta 1$ included a move-up feature to reduce the reservation wait time when a job that was completed before the declared time was removed from the reservation table. However, the series of processes required to maintain coherence while exchanging the many messages involved in executing the move-up feature-messages such as telling the system to cancel the reservation, to suspend the previous queue prior to canceling the reservation, to resume executing reservations, and to resume queue operations-was difficult to verify operationally. The move-up feature was therefore disabled in later versions.

4 Features of NEC's NQS-II/JobManipulator Local Scheduler and Its Use at the Cybermedia Center of Osaka University

In 2001, when the Cybermedia Center of Osaka University adopted NEC's SX-5 supercomputer, the center also installed NEC's NQS-II POSIX-compliant batch job scheduler. The NQS-II includes an optional ERS (enhanced resource scheduler) feature, which can be used to enable process migration between nodes, making it possible to balance the processing load and at the same time achieve flat-rate usability through fair-share scheduling. However, with the NQS-II, although the operating efficiency was high, there was criticism from users that the execution time was uncertain.

In 2007, the center upgraded its system to the SX-8R and installed JobManipulator, a new optional feature for controlling several hundred nodes, originally developed by NEC for the Earth Simulator. JobManipulator uses fair-share scheduling to balance the processing load of jobs on the reservation map before the jobs are executed and suppresses job migration during job execution. This gives users advanced notice of factors such as the length of time until a job is executed and when execution will finish, without sacrificing operational efficiency.

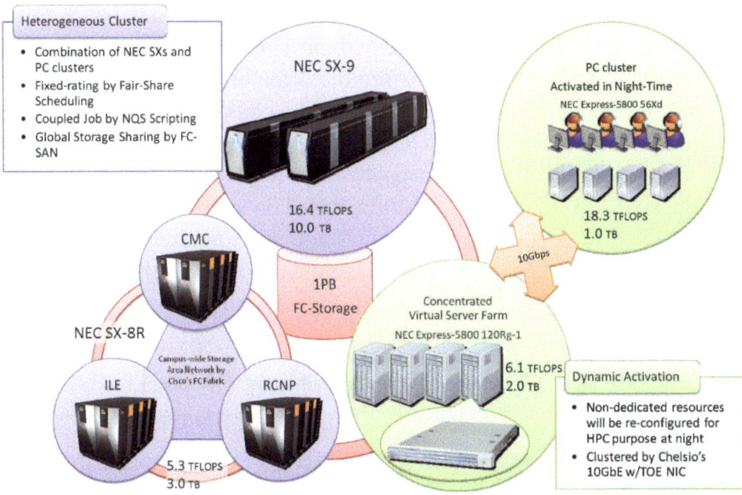

Fig. 2 System configuration of the Cybermedia Center

For the Earth Simulator, resources were managed in node units, but in the commercial version of JobManipulator, resources can be managed in CPU units and according to memory space. Resources can also be diverted to PC clusters. NQS-II is used in all areas of the center's heterogeneous environment so that jobs can be input using the same procedure regardless of whether the system is the SX-9, the SX-8R, which is installed in various locations on campus, an exclusively used PC cluster, or a PC cluster that can be appropriated when idle (see Fig. 2).

Moreover, as we reported last year, by using the NAREGI grid middleware within this framework, the center could realize an environment in which resources are supplied to a grid, thereby enabling the use of all computing resources via a single sign-on.

5 GridVM for SX

In the previous sections, we described how we reviewed the design of the GridVM and added features so that problems with the NAREGI grid middleware could be resolved. We also looked at how the features of the NQS-II and JobManipulator could be used to manage the resources of a vector supercomputer more efficiently.

5.1 Creating a System to Verify the Coherence and Consistency of Web Services

Before starting development, we implemented a time capsulization feature that enables the message exchanges between the Super Scheduler and the GridVM, as well as the associated status changes that occurred in the Super Scheduler, to be captured chronologically. We also implemented an Ajax-based Web user interface to trace the captured time capsules. By using these features, message exchanges can be monitored in real time, and sometimes retrospectively, providing a verification environment in which debugging can be performed based on real-time or retrospective status changes. With the time capsulization feature, the SS/RCS reservation table can be visualized, enabling verification based on visual inspection (see Fig. 3). By cross-checking the captured time capsules with a reservation table displayed using an existing feature of the NQS-II/JobManipulator (see Fig. 4), we were able to verify the consistency of our research results with a high degree of accuracy.

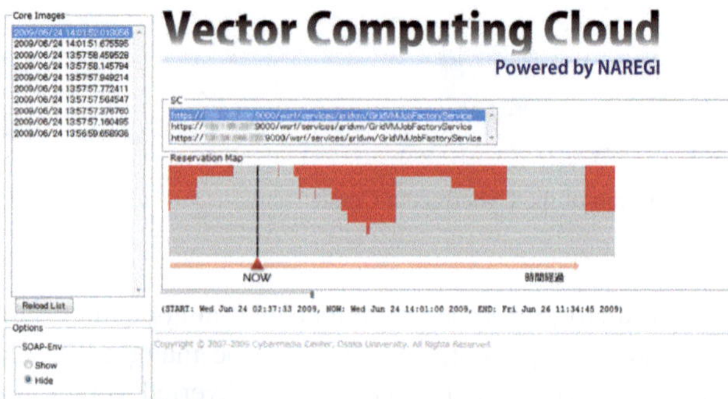

Fig. 3 Visualizing a reservation table of the SS (gray: reserved, red: non-reserved)

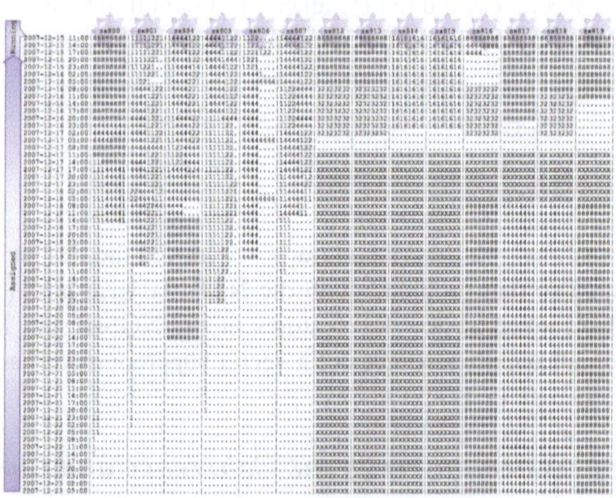

Fig. 4 Visualizing a reservation table of NQS-II/JobManipulator

If we can achieve the same kind of traceability with the OGF grid service-a Web service traditionally difficult to test using integrated evaluation methods-we will be able to shift from supplying a simple service based on loose information exchange, to supplying a complex service based on detailed information exchange; in other words, a service that provides complex processing such as that of the previously mentioned move-up feature.

5.2 Delegation of Reservation Table Administration Authorization by Synchronization of Tables

As described before, the Super Scheduler is designed to realize the adaptability to operations at a computing center through an ad hoc enhancement for non-reservation type jobs, but it can barely allow the user to submit jobs because of the lack of job scheduling capabilities.

As explained in the previous section, fair-share scheduling is realized at the center with the reservation table managed by JobManipulator. If this table can be synchronized with the table administered by SS/RCS and SS can carry out job scheduling by filling out this table, efficient co-existence of jobs is expected to be available in the center operations.

SS/RCS retrieves the information on the time-varying reservation status for each event. Since the reservation table managed by SS/RCS itself loses coherency with GridVM in such an event as system restart, there are functions of preserving the table (inactive as a default) and updating the reservation table on the GridVM side when the revisions of the two reservation tables are found to be different after periodical checking of the tables.

In terms of the latter function, an extended function was developed in order to add to the GridVM reservation table the reservation information related to the jobs reserved by a scheduler other than the Super Scheduler. Furthermore, the event monitoring function was implemented for incorporating the reservation information on center-operated jobs into the GridVM's reservation table from time to time, making it possible to indirectly synchronize the reservation tables of both SS/RCS and JobManipulator. The APIs provided by NQS-II are used for retrieving the reservation information on center-operated jobs, and it was confirmed that any possible inconsistency can be eliminated through the analysis of command execution and log trace, and the coherence needed for operation is secured. By hiding the restrictions arising from the use of the Super Scheduler, the co-existence of reservation-type jobs and NAREGI jobs was made possible.

As described before, one of the issues of NAREGI is the distributed management of SS/ RCS. Here the job scheduling by the Super Scheduler is conducted by retrieving the reservation table that is held by SS/RCS for each GridVM. Therefore, as far as the message exchanges by the Web service are assured, the functions provided by SS/RCS can be delegated to GridVM, thus enabling the simplified entire system configuration and proper load balancing. As a more radical measure, advanced capabilities are needed, such as the inquiries about the status of the reservation table, the re-issuance of request or the penalty for the GridVM that has not responded within a certain period of time, the speculative issuance of reservation, and the cancellation of reservation for improved efficiency.

5.3 Co-existence with GRAM/MDS Interface

NAREGI employs the Globus Toolkit 4-driven Web service as a basis for the software implementation; for example, the GRAM interface is used for job submission. On the other hand, the GT4 binding (GRAM/MDS) for NQS-II has already been developed on a trial basis, and its co-existence with the NAREGI implementation was found to be available. In conjunction with the reservation table synchronization function developed at this time, co-existence of different types of jobs is expected to be available in actual operation, including ordinary jobs, NAREGI-submitted jobs, and Globus-based jobs. We are planning to verify their interoperability through the future operation tests with the aim of flexible operations at the computing center.

5.4 Enabling MPI/SX Job Execution

In the NAREGI project, the execution of the GridVM, which is a concurrent job across multiple clusters, is required. Its use needs the description of applications with GridMPI and GridRPC. Meantime, special inter-node connection devices are used for high performance computing resources, and so broadband, low-latency communications are made available only with the MPI library etc. that refers to vendor-supplied communication libraries. In the NAREGI grid middleware, there is no mechanism for controlling vendor-supplied MPI programs by using such methods as command line options, environment variables, and temporary files, which hampers the dissemination of the NAREGI middleware for supercomputing centers that handle high performance computing resources.

In this research, the activation of vendor-supplied MPI programs was made possible by adding supplementary records to the JSDL generated by the GridVM. We are also investigating the possible submission of MPI-driven jobs across multiple clusters through the reference to the NAREGI-provided information on multiple clusters.

6 Future Issues

In this paper, it was shown that the GridVM for the SX Vector Computer, as the outcome of this research and development, has the potential of enabling the center operations capable of flexibly handling NAREGI workflow-driven applications and Globus application-enabled grid type jobs. NEC's NQS-II already has its implementation for PC clusters, allowing the user to manage and control such PC clusters with almost the same capabilities as those for the SX. In addition to the existing NQS script language for easy description of coupling simulations, the framework shown in this research will enable the flexible manipulation of heterogeneous computing environments.

The NAREGI Super Scheduler has still maintained its presence as a unique metascheduler capable of running reservation-type jobs by synchronizing multiple computing resources, and further continuous efforts are expected to resolve the issues indicated in this paper. In particular for non-reserved type job scheduling, further advancements of the functions are essential in center operations, such as the job scheduling taking into account load averaging and mean queuing time, as well as fair-share scheduling. Prompt complementary measures are needed, including the porting of proven job scheduling functions.

Further deployment of the NAREGI middleware will contribute to the spread of authentication and authorization infrastructures indispensable to building grid environments. Promotion of a mashup for secure Web services is expected to enhance productivity, thus facilitating the rapid establishment of service-oriented grid environments.

References

1. NAREGI. http://www.naregi.org/
2. Sakane, E., Higashida, M., Shimojo, S.: An application of the NAREGI grid middleware to a nationwide joint-use environment for computing. In: High Performance Computing on Vector Systems 2008, pp. 55–64 (2008)
3. Sakane, E., Higashida, M., Okamura, S., Akiyama, T., Baba, K., Shimojo, S., Teranishi, Y.: An authentication system development in application of grid middleware to a nationwide joint-use environment for computing. In: Proceedings of IEEE/IPSJ Symposium on Applications and the Internet Workshop (SAINT2008), pp. 277–280 (2008)

III
Applications

From Static Domains to Graph Decomposition for Heterogeneous Cluster Programming

C.B. Navarrete, E. Anguiano

Abstract In this paper we present the development of two finished projects that have given us enough knowledge to begin with the implementation of a middleware application based on the MPI extension. This ongoing project is also presented in this document. This middleware allow us to dynamically generate a set of domains that fits to the problem to be solved on an heterogeneous cluster. The main goal is to adapt the partitioning to dynamic and unpredictable load changes on the nodes. We want to adapt the architecture to the problem and not the problem to the architecture.

1 Introduction

In the last years, the computers have increased their capacity of calculus and networks—for the interconnection of these machines—have been improved until obtaining the actual high rates of data transferring. The programs that nowadays try to take advantage of these new technologies, cannot be written using the traditional techniques of programming, since most of the algorithms were designed for being executed in only one processor, in a nonconcurrent form, instead of being executed concurrently in a set of processors, working and communicating through a network. This is one of the things we observed as we were implementing the three algorithms presented at this work. The three problems are introduced following a natural improvement of the performance of the solutions from the point of view of the loadbalancing and the complexity of the algorithm used to part the problem into domains.

C.B. Navarrete
Escuela Politécnica Superior, Universidad Autónoma de Madrid, Madrid, Spain, e-mail: carmen.navarrete@uam.es

E. Anguiano
Centro de Referencia Linux, Universidad Autónoma de Madrid, Madrid, Spain, e-mail: eloy.anguiano@uam.es

M. Resch et al. (eds.), *High Performance Computing on Vector Systems 2009*,
DOI 10.1007/978-3-642-03913-3_11

The first section introduces a physical problem, the epitaxial surface growth. This is the most simple simulation, from the point of view of the parallelism, in comparison with the two following ones. It presents a very complex numerical model with a hard requirements of memory and computational power. This first work introduces some initial ideas and concepts required to define the final dynamic domain decomposition.

The second section of this work presents a more relevant parallelisation but also easier numerical algorithm, the Potts simulation. At this section, some different domain decomposition are explained, making a comparison between all of them and concluding with a relationship between the domains used and the performance achieved. This work was usefull to define how groups of processors must be done to obtain the best performance, applying this to a particular physical problem.

Last section joins the results of both simulations previously explained and introducing graph and optimisation concepts describes how a middleware to dynamically generate domain decompositions can be implemented, taking into account characteristics of the computation nodes for an automatic workbalancer allowing to modify data distribution and the network communication topology This partitioning is then distributed onto heterogeneous computing resources, described using the MPI_Graph topology communicators.

2 Epitaxial Surface Growth

Usually, theories of surface growth are based on the study of global processes without taking in account the local behaviour of atoms. In this work we present two simulations making use of a parallel computing library. These two simulations are based on a simple model that allows us to simulate the surface growing process of a certain material. The first one is a quasi-static model whereas the second recreates the atomic interaction considering the free atoms in continuous movement along the surface. Both simulations makes use of local principles of thermodynamic for atomic deposition, relaxation and diffusion of a growing surface. The obtained results agree with those that use global theories and with experimental results of Scanning Tunneling Microscopy (STM).

2.1 Introduction to Physical Model

Surface growth of materials is a very well studied topic from the theoretical and experimental point of view [1–6]. The surface structure of certain materials presents a granular morphology randomly distributed along the surface, with hills of various heights and lateral extensions (structure typically named as mounds) [7]. This property as the roughness (with less significance) are properties which behaviour has been defined as fractal in several times [8–10].

How films growth involves two sciences: the thermodynamics and the kinetics. The kinetics appears because of the continuous movement of the atoms of the surfaces due to the presence of thermal, kinetics and chaotic fluctuations [11]. The thermodynamics appears giving the conditions of nonbalance due to the temperature of the substratum. Calculating those parameters we can grow a surface initially layer by layer until complete the first ones and growing then the surface making mounds. In this case, the growth process is named as epitaxial growth [12].

2.2 Simulation

We want to simulate the epitaxial surface growth of materials taking into account the atomic interaction between free atoms that are in continuous movement along the surface. This is a Laplacian tightly coupled problem with a very high requirements of memory and computations.

The two simulation we have implemented, use two Montecarlo methods: one at the spatial-temporal dimension, determined by the deposition flux, and related to the temperature of the substratum, and another at the spatial dimension also related with the temperature.

These simulations [13] are both based on local behavior, calculating the partition function of the accessible states of the atom (Fig. 2). This allows each atom to know its probability of diffusion towards the next positions. One by one, these atoms will randomly choose, based on these probabilities, the direction in which finally will move. As it is a stochastic process, the direction in which each movement will finally be, not always has to be the most favorable for optimize the total energy of the substratum. Once all the atoms have been diffunded until their positions, the deposition process will start again, repeating the algorithm explained (Fig. 1), a fixed number of times, according to the temperature of the growing interface.

The parallel version uses the LAM-MPI API [14–16]. The main difference between this algorithm and the non-parallel one is that this one divides the computation in several processes. These processes could be of two different kinds: the slaves that simulate the deposition of atoms, the calculus of the energies contributed by the neighboring atoms, the diffusion of the atom on the surface and the simulation of the interaction; the master only manages the synchronization of the different slaves processes. This election has been adopted in order to clearly separate the communication between MPI processes, the synchronization with all the nodes of the cluster and the algorithm of the problem to resolve. To implement the communication between all processes we have study several structures of queues that also permit us to study different models of atomic interactions in the growing surfaces.

```
begin
    foreach layer ← 0 to MONOLAYERS do
        foreach deposited_atom ← 0 to DEPOSITIONS do
            Random_Deposition
            foreach diffunded_atom ← 0 to ATOMS do
                probability ← Calculate_Diffusion_Probability()
                if probability > VALUE then
                    /* the atom diffunds                                  */
                    energy ← Calculate_Energy()
                    Calculate_Direction(energy)
                else
                    Move_Atom()
                end
            end
        end
    end
end
```

Fig. 1 Pseudocode of the epitaxial quasi-static surface growth

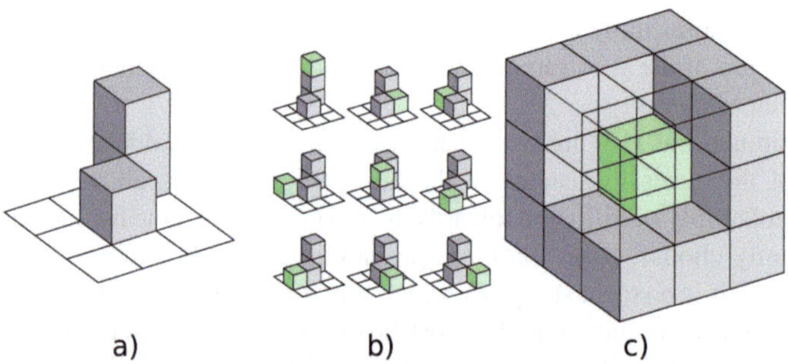

Fig. 2 (a) Example of the surface in a certain moment of the simulation. (b) Accessible states of one deposited atom. Elemental diffusions that can be done by an atom centered in a 3×3 matrix based on the previous surface. (c) First neighbors area that contributes with energy to the central atom. Calculus of the global energy of an atom

2.3 Domain Decomposition

Different computations have been performed: the sequential (a quasi-static model with no possible atomic interaction) and the parallel simulations. For the parallel one, the domain decomposition depends so much, not in the physical result of the execution (since this one only depends on the material and on the distribution of randomness of positions) although it is related to the optimization and speed of processing. Communications between processes spend time that physically can be considered as a variation of temperature. Thus is very important to get a good domain decomposition that optimizes the communications between processes. The domain decomposition consists on a distributed or on a shared solution. We have study diverse domain decompositions that can be resumed in these ones:

• **Partitioning the image in rectangles:** a distributed solution in which the problem is partitioned in horizontal or vertical stripes depending on the number of slaves, value known as N. The most important advantage of this method is the simplicity of the management of the data from the master node to the slaves. The main disadvantage of this decomposition resides in that it implies different rate of growths between subdomains due to the differences between the speed of the processors. Also we have to get in mind that atoms are in continuous movement along the surface and nothing can assure that any atom would not leave its subdomain in its movement, if it was deposited at another subdomain, that could not be in the same slave. Due to this computation model the interactions among processes are to intensive spending much more time in the construction of the communications messages than in the simulation (Fig. 3).

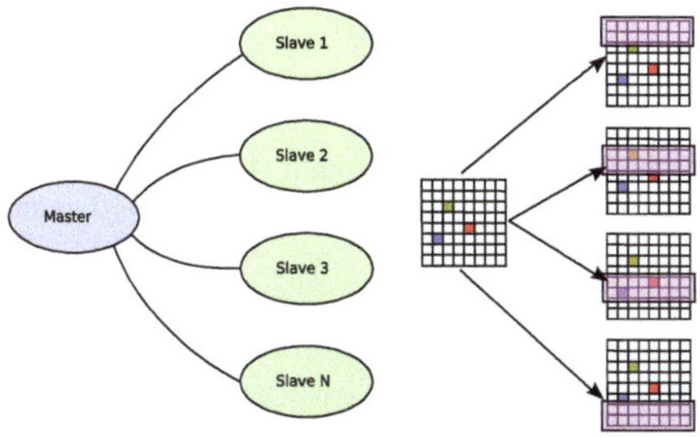

Fig. 3 (1) Image at the master process. (2) Domain decomposition for the problem with N slave processes

• **Partitioning the image according to the environment of the atom:** we define the environment of an atom as a matrix of 5×5 elements that contains the information about the first neighbors and about the second ones (distance of 2 units on each direction) of a certain atom. In this mixed distributed and shared solution, the environments of the atoms are calculated dynamically by the master process and this is the only information that the master sends to the slave. Thus, this method saves much memory in the slaves. The disadvantage resides in that this partition of the image requires too much time and the real computation on each step is so short that it is hard to get good speedup (Fig. 4).

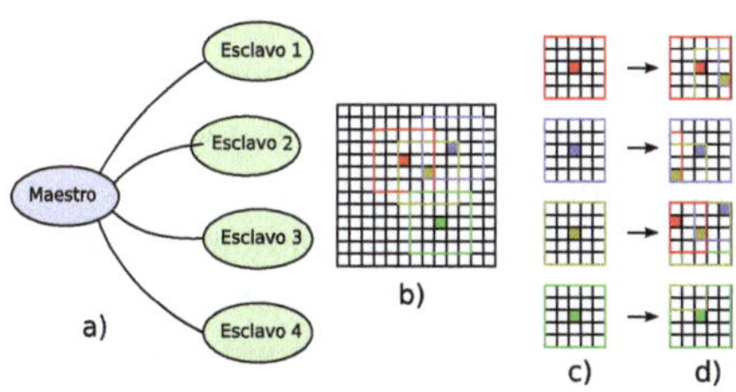

Fig. 4 (1) Image at the master and decomposition in environments. (2) 5 × 5 matrix and environments at the slaves

- **Replicating:** shared solution in which all slaves have the subdomain replicated from the master process. The master and the slaves have shared the data structure that represents the surface. This allows the slaves to calculate the bonding energy of an atom without knowing what have calculated any other node of the cluster. The main disadvantage is that all slaves have to allocate the part of the surface in memory (Fig. 5).

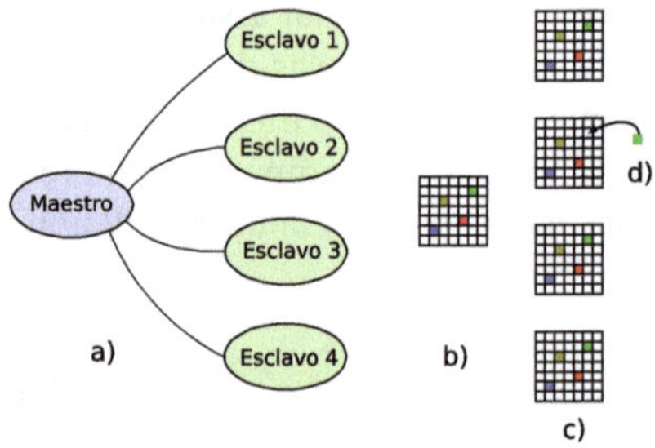

Fig. 5 (1) Calculus of the final position of the atom by the slave. (2) The slave send to the master that position. (3) The master notifies the position to all slaves inserting the result in each queue

2.4 Atomic Interaction

To simulate the atomic interaction between atoms in diffusion, we need to use a queue structure. The atomic interaction and the deposition flux depends on the queues and its treatment. As we have seen, the queues are essential to discouple the two Montecarlo methods that are involved in the solution of this physical problem. We have performed several test of random number generation with the SPRNG Library [17] to probe that the Montecarlo Methods from the point of view of the random number generation were totally discoupled. Results are not showing any measurable difference between whether using this library. The main difference between this two methods is the execution time spent. This is the reason why we use the less expensive computationally algorithm that is the simple random number generation.

The queue structure (Fig. 6) consists on a double data array managed on the master process. The data structure allocated in the queue consists on two data structures. Each structure allocates two coordinates used to identify the position of an atom at the image. The two points represent the movement of the atom on its diffusion process, from the first position towards the second one. This structure, also contains a counter that defines the occupation of the queue. This variable is needed for readjusting the tasks for each process allowing us not to have to consider the speed of processing of each nodes of the cluster.

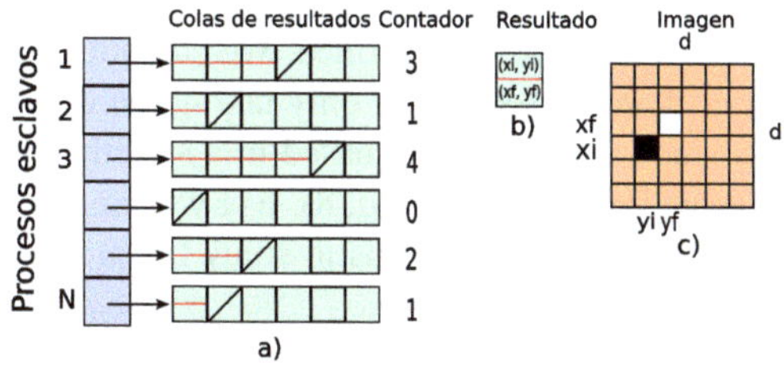

Fig. 6 Queue structure at the master process. (1) Double array data structure. (2) Entries counter for readjust the tasks of the slaves. (3) Data structure allocated on the queue. (4) Representation of the movement of the atom according to the information shown in the data structure

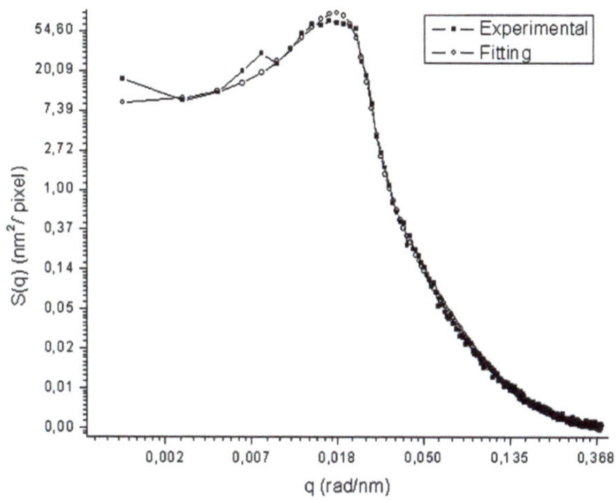

Fig. 7 Fitting between the experimental and the simulated results

2.5 Results

Analyzing the results of these domains decomposition, we have chosen the pure replication solution, just because this is the one that better fits to the problem of the atomic interaction, from the point of view of the optimization and speed of processing, due to the minimization of the communications between the master and the slaves processes.

In order to verify the accuracy of the simulator, the results obtained with existing results were compared with STM experiments results [7]. This analysis verify that the results are similar to the ones obtained experimentally, as it can be seen in Fig. 7.

From the point of view of the execution time results we can observe that the non-parallel algorithm spends much time even more if we try to use it to recreate the atomic interaction. The results of the parallel version have been obtained executing the application in a 8-MPI heterogeneous NUMA [18] LAM-MPI [14–16] cluster. 4 of those 8 processors were a simple Pentium–III Clamath and the other 4 were the processors of an IBM e-Server SMP–Xeon.

Different computations have been performed. We can not conclude that the parallel algorithm is better than the sequential only taking the execution time. This is an unfair comparison since algorithm 2 is based in a slightly different model as it regards additional physical interactions (Fig. 8). But we can conclude that the parallel algorithm is better because it permit us to study the atomic interaction in the growing surfaces in a reasonable time.

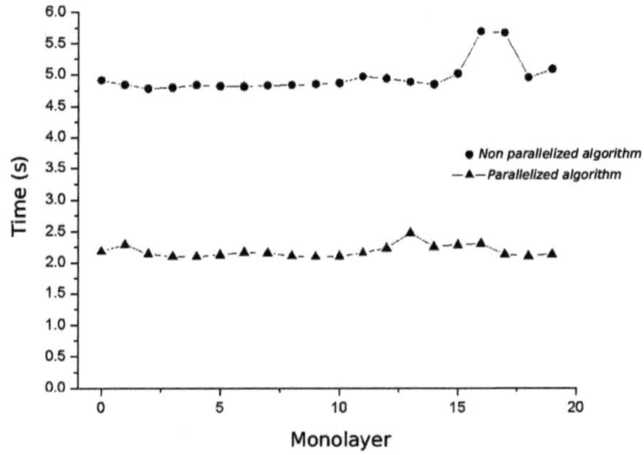

Fig. 8 Comparison of the execution time between the two developed simulators. The growth parameters were: energy $\epsilon = e/k_bT = 0.5$ and dimension $d = 32$ pixels^2

3 Potts Model Simulations

The main idea focuses on the simulation of the Potts Ferromagnetic model, by the allocation of resources and tasks and the configuration of the topology of the network, by transferring the workload onto other computers of the farm, to find dynamic matching between the tasks and the global resources of the NOWs, that optimizes this simulation completion time. This will be done assuming a non-static and decentralized approach.

In recent years, many different single-cell-based models have been developed and applied successfully to various biological and medical problems. These models use different computational approaches [19]. One of these models, developed by Graner and Glazier [20], is called the Cellular Potts Model (CPM), an extension of large-q Potts model, used to study differential adhesion and cell sorting in confluent sheets of cells. Some recent applications of the CPM can be seen in [21, 22].

This model has the ability to model thousands of individual cells in space. Each cell has its own set of properties, such us size, type or adhesion strength to neighboring cells.

Cells are represented as extended objects on a square lattice and are assigned a unique number (called spin in the large q-Potts model) and a type. Adhesion between cell types is incorporated by defining surface energies between neighboring cell membranes.

The central component of the CPM is the definition of the Hamiltonian, determined by the configuration of the cell lattice. The lattice is usually defined as a two-dimensional rectangular Euclidean lattice, but it is often generalized to other dimensions or lattices. The Hamiltonian is defined as follows:

$$E = \sum_{x,y} J_{\tau(\sigma(x)),\tau(\sigma(y))}(1 - \delta_{\sigma(x),\sigma(y)}) + \lambda_\nu(\nu_\sigma - V_\sigma)^2 \quad (1)$$

where ν_σ is the volume in lattice sites of cell σ, V_σ its target volume and λ_ν the strength of volume constraint; $\sigma(x)$, $\sigma(y)$ is the spin at site (x, y), J is the interaction energy between neighboring sites labeled by $\tau(\sigma(x))$ and $\tau(\sigma(y))$ and τ denote cell types.

The evolution of the model proceeds using the Metropolis Montecarlo [23] simulation as follows:

1. Randomly select a lattice spin
2. Randomly choose a neighboring spin of the current site
3. If the neighboring spin is equal to the spin of the current site, go to step 1
4. Change the current site spin for the spin of its neighboring site, according to the following probability function:

$$P(\Delta E) = 1 \quad \text{when } \Delta E \leq 0$$
$$P(\Delta E) = e^{-\Delta E/k_B T} \quad \text{when } \Delta E > 0$$

   where k_B is the Boltzmann constant and T is a certain temperature value.
5. Return to step 1

The lattice can be represented as a torus and is continually updated: for each lattice point, a different spin state is proposed and the new overall energy calculated. The energy depends on the interactions between neighbors and the overall temperature of the lattice. If the new energy is smaller than the old one, the new state is accepted. If not, there is still a certain chance to be accepted, leading to random spin flips which represent the overall temperature.

The critical part of the algorithm, from the point of view of parallelisation, resides in steps 2 and 4. The domain and data decomposition does not always assure us that the variables needed for the computation are known by the processor that calculates a certain value.

Our work describes an alternative parallel CPM solution, more efficient than the sequential method, which uses a C and an MPI library.

The parallelisation of CPM is an active area of research [19]. There are several references about parallel implementations of the CPM model, such as those described in [24, 25]. Our goal is to parallelize the CPM to be executed in a parallel environment, created by an heterogeneous cluster and using the MPI library [16].

We trend to increase the speed and efficiency and reduce the drawbacks, assuring that the required values for the spin-flip attempt computation are accessible by the processor which is computing the possible spin-flip. On the other hand, after a Monte-Carlo step (MCS) the domain decomposition must always be correct.

Our work describes a parallel Montecarlo solution, using a C and an MPI library [16], which is more efficient than the traditional methods. Our goal is to parallelize the Potts model to be executed on a parallel environment, created by an heterogeneous cluster, using the MPI library. The aim is to increase the speed up and the efficiency compared to the traditional Potts simulation. There are several publications about parallel implementations of this model such us the one described in [26]. Many parallel algorithms have recently been developed to reduce this "drawback" [26, 27]. The problem of these algorithms is that they are much more difficult to parallelise efficiently. These processes are the result of applying a division method to the domain of the problem.

The parallel version uses the LAM-MPI API [14, 15]. The main difference between this algorithm and the non-parallel one is that this one divides the computation in several processes. These processes could be of two different kinds: the slaves, that solve the problem and the master that only manages the synchronization of the different slaves processes. The communications between the master and the slaves processes is made by the use of signals and tags to identify them.

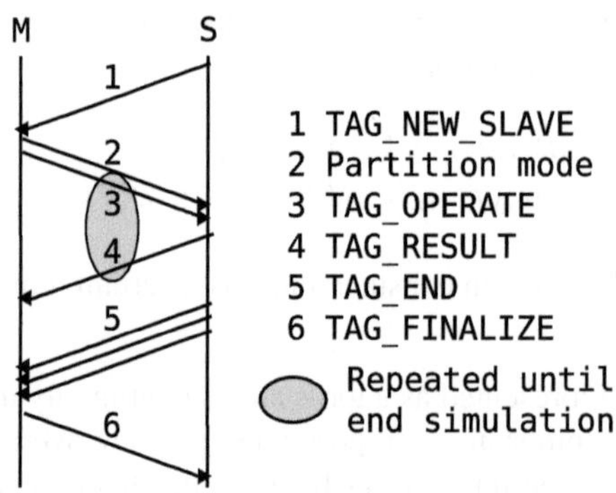

Fig. 9 Scheduler with the signals used in the parallel implementation to simulate the Potts model

The program is divided in 3 sections (see Fig. 9). The first one, in which the slaves join the simulation, is defined by the tag TAG_NEW_SLAVE. This first section of the simulation is very important to calculate the number of processes that have joined the simulation, in order to define the properties of the domain decomposition selected to be used. The second part of the simulation solves the Potts simulation, depending on the chosen domain decomposition. This domain decomposition is sent to the slaves processes identified by a special tag. The elapsed execution time is calculated between the tags TAG_OPERATE and TAG_RESULT in order to avoid the spent time to allocate the memory for the buffers, to restore and image and to save it, etc. This part is repeated until the simulation finish, that is, when a certain number of monolayers in the simulation process have been completed. The final part of the simulation, is needed to detect the end of the processes, free the allocated memory and finalize the execution of the MPI virtual machines.

3.1 Domain Decomposition

The critical part of the algorithm, from the point of view of the parallelisation, resides in the step 2. Not always the domain and data decomposition assure us, that the needed variables for the calculus are at the processor that is calculating a certain expression (see Fig. 10).

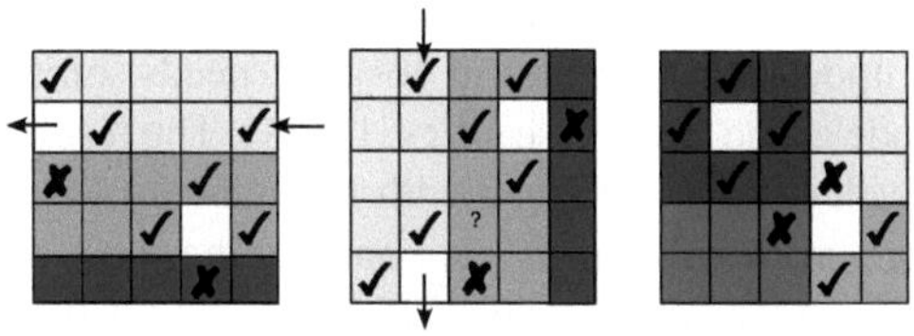

Fig. 10 Depending on the domain decomposition, it can happens that the neighbors of a certain position of the grid are not placed in the same processor. In this situation, it will be necessary to communicate with the neighboring processors

Five simulations have been performed: one simulation for the non-parallel applications and four more to study the best domain decomposition, in order to optimize the communications between pairs of processes, to increase the speed up of the execution of the problem. In these simulations we have decided to use a domain decomposition based on horizontal or vertical stripes; on a square decomposition; or based on a domain decomposition based on the neighbours grid position of the data grid (environment decomposition). In the last simulation we have decided not to use a domain decomposition, using queues to communicate nodes. These initial domains decomposition are needed to get the most performance values as possible, in order to study the behaviour of the problem, depending on the initial input values, the initial distribution of nodes and data. The goal of our algorithm is the way we solve the possible situation in which one node needs some information, in order to update the current grid position that is in another node of the cluster, due to the domain and data decomposition.

If we divide the problem into stripes, the image is partitioned in horizontal or vertical parts obtaining a total number of N subimages. The main goal of this domain decomposition is that it is a very simple domain to recreate because of the continuous memory architecture of a PC. On the other hand, there are defined long borders area shared by each pair of neighboring processors, which implies a high probability of communications between these 2 processors. Another great disadvantage is that the slave must allocate too much memory, to receive the corresponding subdomain. At the end of the simulation, It is slow to send the results to the master. Dividing the problem into squares is very similar to stripes decomposition, but in this case, the subdomain follows the geometry of the initial lattice. It takes into account the symmetry of the problem. The ratio of communications between pair of slave processors is negligible when the size of the image is much greater than 0 ($d \gg 0$). The main observation to use this domain decomposition resides in that the number of processes must be a square number. As in the stripe domain decomposition, the slaves must allocate much memory to receive and manage the subdomain. Dividing the problem according to the environment of the spin of the grid needs a buffer of 5 elements that contains the information about the first order neighborhood. This neighborhood is needed to calculate the energy of each spin at the surface. The main advantage of this simulation is that the environment is the only information that the master sends to the slave, so it is very fast to send the results to the master. It saves too much memory at the slaves but a deadlock between neighbors could exist.

The goal of our algorithm is the way we solve the possible collision, situation in which one node, to update the current grid position, needs some information that is placed in another node of the cluster (see Figs. 10, 11). The results obtained with this simulation are equivalent those obtained with the standard Metropolis Montecarlo algorithm. The future work will consist on applying these concepts to simulate the cell growth using a Cellular Potts Model (CPM [20]) and using a dynamic domain decomposition based on the cell dimensions.

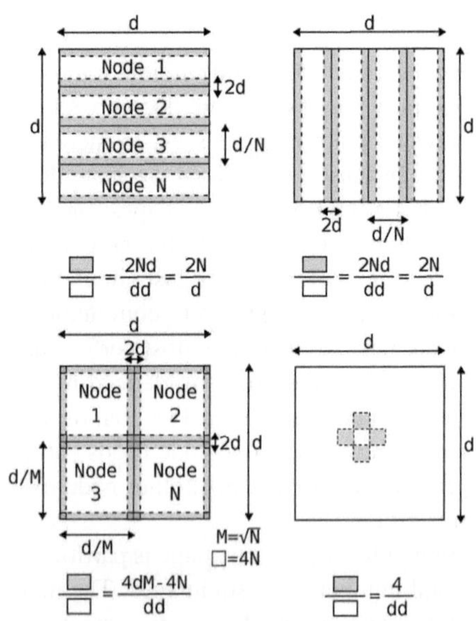

Fig. 11 Graphical explanation and mathematical expressions that define the concept of the border between pair of processors. When a pixel of the border is randomly selected, the process must send a special signal to the master process to get the value of the pixel that is not in its subimage

3.2 Results

The results obtained with this simulation are physically equivalent to ones obtained with the standard Metropolis Montecarlo algorithm (non-parallelised simulation). This assure us that the initial partition has nothing to do with the Potts model and the Ferromagnetic properties from the point of view of the physical results.

We have compared the results of the parallel version with those obtained with the non-parallel one of the same problem. We have considered, for all the simulations, a toroidal Potts lattice; that means that if we select a position from the lattice, its neighbourhood can be out of the simulation surface limits, but in our case, we have considered that one atom situated on the limit of the simulation area, has its neighbours at the opposite limit. This is only applicable to the initial lattice and it is not true for the domains that the slaves receive. For these processes, a spin located at the border area of each subdomain (corresponding to each slave) has its neighbours located at another processor, but not at the opposite limit of its lattice. To get the value of the spin that is in another processor, the current slave must send a notification (communication) to the master process in order to ask for this value and calculate the probability partition function for this current grid position. The master will ask the processor that owns the pixel to answer the petition of the first slave. This protocol between processors increase the computational time to solve the Potts problem. This increase of the execution time is related to the probability of choosing a spin of the neighboring area, that is related to the rate between the perimeter of the border and the total area of the subdomain of the slave process (see Fig. 11) and of course, related to the domain decomposition chosen for the simulation.

If we simulate the Potts model using a decomposition based on stripes (see Fig. 12), we can see that the problem is not scaling. We can see that time decrease a little bit if we add a second processor to the simulation but,as we continue adding new processes to the simulation, the time continues increasing. This happens because of the great border area in comparison with the total area of the subdomain received by each slave. We are spending more time creating the network packages than solving the simulation. To get an easier comparison between the results, we have added a line to the graphics, that represents the ideal scaling between the number of processors and execution time. The main idea of the ideal scaling is that, if we multiply by two the number of processors, time should be reduced by half. This is not completely true because we must also add the time for the creation of the network packages and the time for sending them.

We can see (Fig. 13) the same problem solved using a squares decomposition. In this case, we observe that the problem is scaling with the size of the problem, and it is very near to the ideal behaviour of the simulation. For small lattices, we can not add more slaves processes to the simulation, so we can not have more results to be compared with the simulation of the Potts model, using images of bigger sizes. We can also observe that for each line of the graphic, we can determine a local minimum time value which, according to the increasing of the size of the image, moves to a bigger number of processors. From this point, the problem to simulate is too much divided and the execution time increases as the border area increases. The

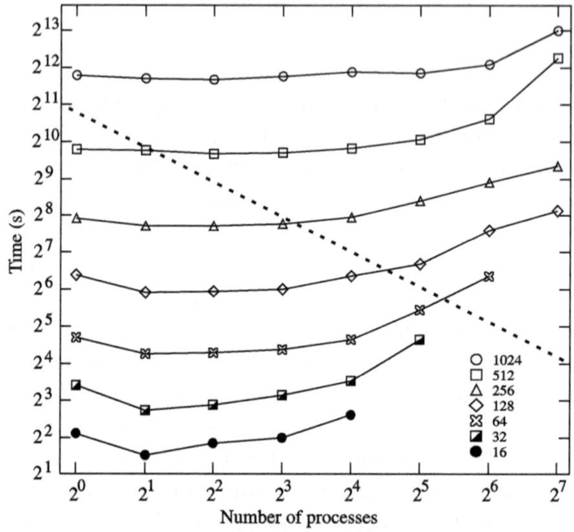

Fig. 12 Execution time versus number of processors graphic, using images from $d = 1024$ px^2 to $d = 16$ px^2. The ideal scaling line is shown. The domain decomposition used is the stripes decomposition. The time is a logarithmic axis

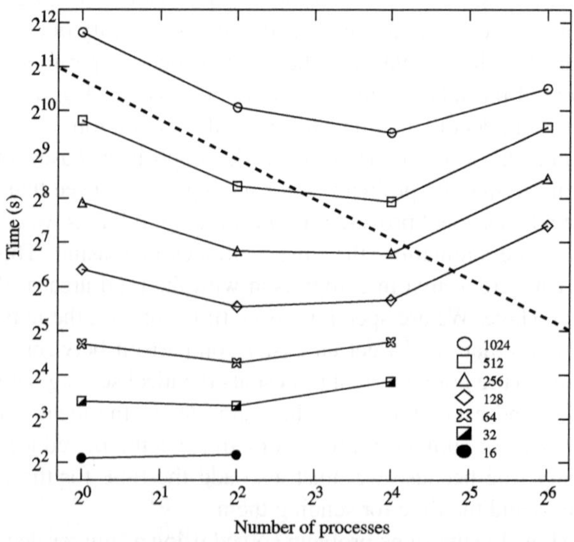

Fig. 13 Execution time versus number of processors graphic, using images from $d = 1024$ px^2 to $d = 16$ px^2. The ideal scaling line is shown. The domain decomposition used is the squares decomposition. The time is a logarithmic axis

time spent in the calculus of the Potts functions is negligible, in comparison with the time elapsed to fill the network packages and send them to other slaves joined to the simulation. In the stripe domain decomposition this minimum value does not exist, due to the scalability of the problem.

Comparing both graphics, we can observe that both have approximately the same behaviour: time decreases from the monoprocess simulation to the execution of the simulation in a cluster, with few processes. Then, we have a more or less constant execution time until this value increases, due to the size of the domain partition and the number of processes, in which it is only important the network transfer time. But we can see, comparing both behaviours, that it is very important to determine the best domain decomposition depending on the problem to be simulated.

4 Graph Domain Decomposition

The use of clustering computing to solve computational problems has been the focus of high-performance computing community for more than two decades. For general optimisation of hosts in farms of computers, a methodology of dynamic resource allocation is needed [28–31].

Homogeneous partitioning methods for N-dimensional data-parallel iterative programs have to deal with several conflicting goals. Workload has to be equally distributed among the processes, communication amount which depends on the surface of each domain and the number of neighbours should be minimized, and the amount of data exchange for re-balancing should be minimized. There are several approaches for homogeneous dynamic partitioning. In all dynamic methods, the partition is changed withing certain re-balancing intervals. The length of these intervals depends on the problem to be solved and some system properties. The length of the re-balancing intervals can be changed dynamically. At the end of the re-balancing interval, a new partitioning has to be computed using information about the workload distribution. Then the processes exchange parts of their local data according to the new partitioning.

There are several methods for homogeneous partitioning e.g [32]:

- Square decomposition: the data is split into a fixed number of sized boxes. This partition minimizes the communications amount between partitions.
- Homogeneous stripe decomposition: divides the data space into rectangular stripes of equal workload. To determine the frontiers or borders of each stripe, some global information about the global workbalance is needed. This method is a centralized parallel decomposition method. The algorithm to solve the frontiers is very easy to compute but an algorithm to synchronize the values of the frontiers shared by more than one processor is needed.

In heterogeneous clusters, all processes should execute their computations steps in the same interval of time. For this issue, we must define the speed of the process as the amount of work that can be done in a certain interval of time [33]. To consider this speed, the possible bottlenecks or net latencies, for example, must be taken into account. This speed can be estimated by some heuristic functions based on other characteristics of the processors of the cluster. Thus, the partitioning method has to adapt the size of each decomposed domain to the speed of the processes involved in the heterogeneous cluster. The goal of the heterogeneous partitioning is to assign the workload to each process according to its individual process weight, which is a normalized measurement related to its speed.

As the homogeneous partitioning can be seen as a special case of heterogeneous methods when the factor of the speed is $1/p$ being p the number of processes, the homogeneous methods could be also applied for the heterogeneous partitioning, but in this case, the efficiency is not going to be desirable due to the complex algorithms needed to synchronize the values of the frontiers that exist between domains. This methods to synchronize the values of the frontiers, must also take into account the possible disparity of characteristics of the processors of the heterogeneous cluster. In addition, not always the regular domain decomposition are the best methods to solve data-parallel iterative problems, but other problems need to be solved, to save resources or due to asymmetries, by using irregular partitions, and thus, the homogeneous methods can not be applied.

Our goal is to generate a dynamic and domain decomposition based on irregular partitioning, that fits to the problem and fulfill with the requirements of the problem to be solved and the characteristics of the nodes of the heterogeneous cluster.

4.1 Model

The HNoC system can be modelled as a weighted directed graph G_s, denoted by $G_s(P, L, \tau, \delta)$, referred as the $SystemGraph$; P denotes a finite set of processors that represents the nodes or vertexes of the graph G_s; L is a finite set of links that represents the communication links between pair of processors: the edges of the graph G_s; Each vertex $p_i \in P$ is characterised by a set of system parameters (memory, frequency, operating system...), based on its available resources of the HNoC. Due to this, each processor has a processing weight $\tau(p_i)$ that denotes the processing cost per unit of computation. Each link between two processors p_i and p_j, denoted by $l_{i,j} \in L$, has a link weight $\delta_{i,j}$ that means the communication latency between those two nodes per transfer unit. If two nodes $(p_i, p_j) \in P$ are not connected to each other, then $l_{i,j} = \infty$. We assume that all nodes of the graph are connected to at least one node of the HNoC (connected graph) but we not enforce constraints on the network topology, as this is not completely defined and can vary between two steps of the simulation of the problem, that is being executed in the HNoC. It is necessary to define a neighbourhood function which will return the set of nodes that are linked with any node of the HNoC,

$$\forall p_i \in P : neig(p_i) = \{p_k\} \mid l_{i,k} \neq \infty$$

So, we must define another function $path(p_i, p_j)$, defined as

$$\forall p_i, p_j \in P : path(p_i, p_j) = \{p_k\}^*$$

were p_k^* is a sorted set of nodes were each node either $p_k \in neig(p_i)$ or $p_k \in neig(p_j)$ or $p_k^n \in neig(p_k^{n-1})$. It is necessary to define a latencies matrix C_L containing the network latencies between any two processors, $\forall p_i, \forall p_j \in P, C_{L_{i,j}} = lat(p_i, p_j)$, which will depend also on the physical and data link network layer. For two adjacent nodes $p_i, p_j, \in P$, $C_{L_{i,j}} = lat(p_i, p_j) = \delta_{i,j}$ but if p_i and $p_j \in P$ are not adjacent in the HNoC, the latency will be defined as the sum of the links weights on the shortest path between them,

$$C_{L_{i,j}} = lat(p_i, p_j) = \sum (\delta_{p_k} \mid p_k \in min(path(p_i, p_j))).$$

The matrix C_L could be symmetric or not, since all communications could be different, also in a duplex communication, because of the directed property of the graph.

An heterogeneous network of computer systems (HNoC) [13, 34] can be modelled as a weighted directed graph G_s, where connections have the weight of the edges (the cost of moving directly from one vertex to another), representing the delay of the network and those delays due to overload of the processors in the nodes. Each processor available at the HNoC is identified by a vertex in the graph. The nodes allocate information related to the effective load of the processor, defined as its availability to execute other processes. The nodes also contain a statistic value proportional to the execution time of the processes in the previous steps of the algorithm. Therefore, the edges of the graph represent the connections between the available processors in a certain instant of the algorithm. These connections are statistically weighted as a function of the communication overhead: the time of transmission of the data through the net (which also depends on the network protocol), and the physical layer used for this communication. These two values are important for the calculation of the heuristic function.

In order to build a computing environment for farms of computers, it is also necessary to have an algorithm that requires the ability to predict the performance and the resource consumption of different cluster configurations. This algorithm is called the Resource Management System (RMS). The problem is that it is difficult to predict the computing time of each processor before it receives some arbitrary load. And this will be even more difficult, if we consider the communication topology variation of each computer of the farm. The basic tasks of the RMS is to accept requests for resources made by the applications and allocate them from the poll of resources. This is a slightly approximation of a computer middleware. The RMS uses a predictive estimation based on a mathematical function (heuristic function, Υ) in order to map the tasks of the parallel application to the pool of resources. This heuristic function is the one needed to define the graph partition and assigning workloads to each node. These techniques for predicting the performance of the dynamic system are nowadays based on queueing techniques and/or on historical techniques [35]. Making a comparison between layered queues and the historical

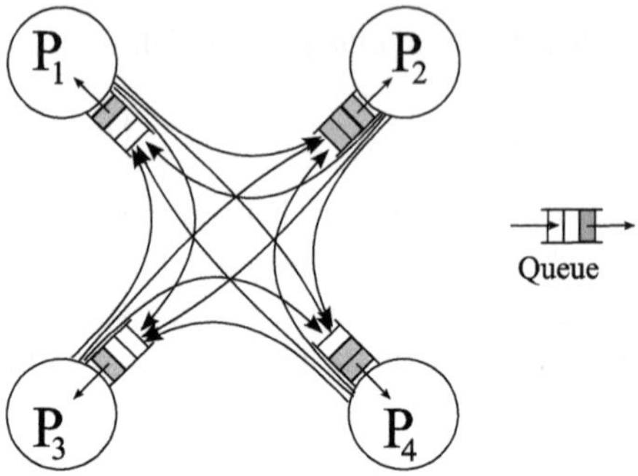

Fig. 14 Example of an HNOC with 4 nodes. Each server has its own queue for receiving the incoming requests from the rests of nodes of the farm

model, it is well known, that the layered queue requires more CPU time to make the mean response time prediction, whereas the historical predictions model are almost instantaneous. However, queuing techniques are easier to implement with a minimum level of performance than the historical model. This is because designing an historical model involves specifying and validating how predictions will be made, whereas the queueing model can be solved automatically. Both techniques can be combined to take advantages of them [36].

The layered queueing performance model defines an application's queueing network. The solution strategy, in this case, involves dividing the problem into tasks depending on the resources and corresponding them to the tiers of servers in the system model, generating an initial topology solution and then iterating with the historical method, solving and resizing the queues in each step of the algorithm, until the solution converge to an optimal distribution of resources, tasks and communications delays.

For the queue model it is necessary to define a queue structure for each node of the HNoC, which will be shared by all the incoming requests (Fig. 14). The nodes can be both slaves (request information) and masters (requests scheduler). The queue can be managed by a FCFS (First-Come, First-Servered), LCFS (Last-Come, First-Servered) or SIRO (Service In Random Order) policy.

The application can also be modeled as a weighted directed graph G_a, denoted by $G_a(T, D, \omega, \lambda)$, referred as the *ApplicationGraph*; T denotes a set of vertexes of the graph that represents the tasks to be done; D represents a finite set of edges of the graph where $\{(t_i, t_j) \mid t_i, t_j \in T\}$; Each vertex has a computation weight $\omega(t_i)$, $\forall t_i \in T$ that represents the amount of computations required by the task t_i to accomplish one step of the algorithm. Each edge has a value $\lambda_{i,j}$ that represent de amount of data to be sent from v_i to v_j. The existence of an edge between vertex A and vertex B, means that, to calculate the value of A at a certain instant of the execution, we need the value of B at the previous step of the algorithm.

Thus, the execution time Γ of a task $t_i \in T$ on a processor $p_j \in P$, assuming the worst case in which there is no-overlapping between computation and communication, is defined as:

$$\Gamma(t_i, p_j) = \omega(t_i) \times \tau(p_j) + \sum_{t_l \in neig(p_k)} \sum_{\substack{p_k \in P \\ k \neq j}} \lambda(t_i, t_l) \times \delta(p_j, p_k) \qquad (2)$$

where $\omega(t_i) \times \tau(p_j)$ represents the amount of computation required by the task t_i per processing cost per unit of computation at processor p_j, the set t_l contains the neighbours of the vertex p_j of the graph G_s, P is the set of vertexes of graph G_s, $\lambda(t_i, t_l)$, represents the amount of data to be sent from t_i to t_l and $\delta(p_j, p_k)$ is the weight of the link between processors p_j and p_k.

Given a system graph G_s and an application graph G_a, our objective is to map their characteristics $\Gamma : (T, D) \mapsto (P, L)$ to minimize function Γ, based on the application requirements and the system constraints, such as the topology of the system graph.

4.2 Workbalance

We want to dynamically generate a graph partition (a set of domains) that covers the data grid and fits optimally to the problem. To achieve this, the size of each generated domain (the number of tasks assigned by the master to each slave process) will depend on the result of an heuristic statistical function (Resource Management System, from now called RMS) applied to some loadable functions that are related to the characteristics of the nodes of the cluster and to the properties of the network connecting them (see Fig. 15). These loadable functions depend both on the physical and net properties of the node and topology, and also on time related properties (historical values).

The RMS value for each processor is calculated by the following expression:

$$RMS_j = \sum_{i \in P_{G_s}} W_i \times F_j$$

where P_{G_s} is the set of vertexes of the graph G_s, and W_i is the normalized weight for each loadable function F_j applied to the vertexes of the set. The initial implemented functions available at the system are:

- Computational power: regards the CPU frequency of each node at the cluster.
- Free memory: takes into account the amount of free memory that the node has.
- Net latency: related to the latency between the node and any other of the cluster. Depends on the network architecture and the topology used.
- Historical values: computes an heuristic function based on the average runtime of the N-previous steps of the algorithm to be solved.
- Time estimation: special case of the previous method if the depth of the historical times is $N = 1$.
- Task complexity: statistical value related to how complex the task is.

It must be assured that

$$\sum_{i \in P_{G_s}} RMS_i = N \times M$$

where $N \times M$ is the total number of vertexes in the G_a, T.

Depending on the problem to be solved and its characteristics, will be desirable to have a set of coefficients to weight W_i the functions for the RMS final value or others.

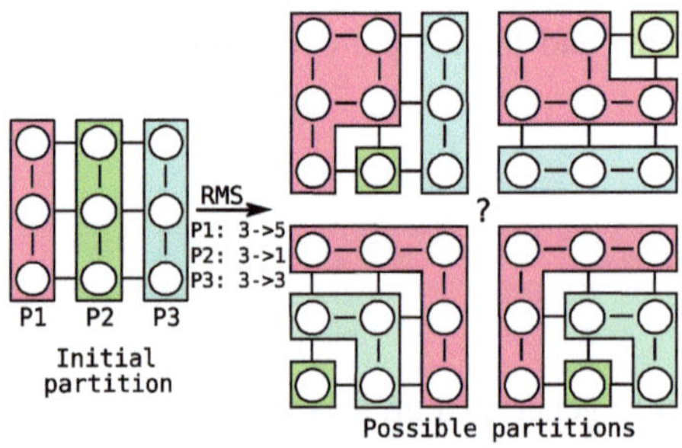

Fig. 15 Starting from an arbitrary data distribution in a 3 × 3 grid graph with 3 processors, P1, P2 and P3, to each processor 3 tasks or values to be calculated are assigned. When they finish, the supermaster node computes the RMS again, obtaining, for example the values 5, 1 and 3, from each process. The procedure to divide the grid into 3 groups with sizes 5, 1 and 3 has not yet been determined and will depend on the symmetries of problem to be solved

4.3 Domain Decomposition. Grouping Algorithm

The RMS value gives us only the number of tasks to be assigned to each processor, but we know nothing about the best data distribution to apply. We determine by the use of a grouping algorithm the assigning of the vertexes of the G_a, in order to define the domain decomposition. The total number of tasks assigned to each processor depends on the RMS value computed by the supermaster node of the HNoC, but the size differences in the domain in different steps of the simulation entail a variation of the data distribution along the cluster. Thus, we have to determine the best data topology for each step of the algorithm, taking into account the data requirements and the symmetries, which also depend on the problem to be solved at the HNoC.

To determine which tasks should be assigned to each processor, we need a searching algorithm that assures that the partition assigned to each processor is the best, from the point of view of the global optimization of the application graph G_a. This algorithm will generate as many groups as slave processes are involved in the simulation, each one being assigned as many tasks as the value of the RMS function returned for the process. Each group contains some vertexes of the G_a and their dependencies.

To determine which processes must be included on each group, a searching algorithm is needed. This searching algorithm can be based on a Breadth First Search (BFS), on a Depth First Search (DFS) or on another search function (also a loadable search method). The accuracy of this solution will depend on this algorithm. If, for a given data partition of the problem, a certain processor does not satisfy its data requirements and need additional data from another processor, it will send a message to the cluster master process (messages between slaves are not allowed), will receive the requested value, and so will be able to continue with the computation. The master will keep this dependency in a local structure. The more messages the slave node sends to the master, the slowest the algorithm and the simulation. Thus, the data grid should be allocated to the nodes of the cluster in such a way that the best performance is achieved. Once the searching algorithm has been executed and the order of the processors has been determined in an array, the domain decomposition is decided by grouping as many tasks from the array as specified by the RMS value for each processor (see Fig. 16). By the use of this grouping algorithm and the information about the dependencies of each data at the partition surface for each process, irregular partitions are possible.

The grouping algorithm will determine the best data topology after each step of the algorithm, taking into account the data dependencies.

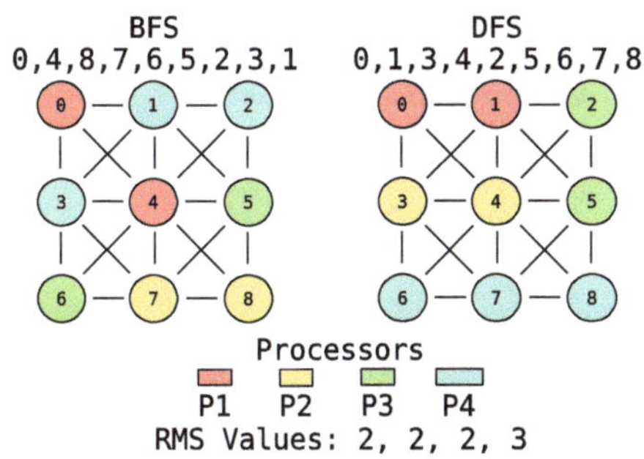

Fig. 16 Comparison between the domain decomposition generated by the BFS and the DFS algorithms. In the example, the G_a contains 3×3 vertexes and the HNoC is made of 4 processors: P1, P2, P3 and P4. In this example, the respective RMS values obtained by the supermaster for each processor were 2, 2, 2 and 3 tasks per processor. The DFS algorithm generates more compact groups than the BFS algorithm, but the latter is fastest, from the point of view of execution time

A better allocation of the data grid into the nodes of the cluster provides a better performance.

4.4 Programming

This application can be considered as a middleware which works between the master (server) and the slaves (clients). It was implemented using the C language with the MPI-2 extension. The master process runs the part of the code that organizes the reconfiguration, the domain decomposition and the optimization of the communication and topology of the HNoC. The slave processes only execute the tasks sent by the master process, according to an RPC function determined by the master process.

The algorithm followed by the master and slave processes is summarized in Figs. 17 and 18.

```
begin
    waitUntillSlavesJoin
    createGraph
    repeat
        calculateRMS
        optimizeGraph
        determineGroups
        sendPartitions
        repeat
            tag ← receive
            if tag! = result then
                receivePetitionNeighbourhood
                sendNeighborToSlave
            else
                receiveResults
            end
        until receiving
        saveImage
    until simulationNotFinished
    killAllSlaves
end
```

Fig. 17 Master process

The System Graph G_s topology is created with the *MPI_Graph_create()* function and managed by the use of the *MPI_Graph_*()* functions. This graph is optimized by a Dijkstra algorithm. Once the graph is optimized, it is time to create and define de groups of communicators. For this issue, the collective communicators are used, and they are created using the *MPI_Comm_group()* and defined with *MPI_Comm_excl()* and *MPI_Comm_incl()* functions. We can manage the communications sending the messages using the specific communication handlers created that corresponds with a partition of the graph and also with a domain decomposition of the data surface of the problem.

4.5 Results

To test this *middleware* it is implemented a test application that consists in a parallel and distributed version of the *Game Of Live* simulation. There is not special interest on this algorithm, only the simplicity of the implementation and the high rate of communication of each cell of the algorithm with its neighbours. The same problem

```
begin
    joinWithMasterProcess
    repeat
        partition ← receivePartition
        executeStepAlgorithm(partition)
        sendResultToMasterProcess
    until simulationFinished
end

/* executeStepAlgorithm procedure                              */
begin
    foreach item ∈ partition do
        foreach neighborItem ∉ partition do
            askNeighborToMaster
            receiveNeighborFromMaster
        end
    end
    /* CPM Algorithm in this case                              */
    executeSimulation
end
```

Fig. 18 Slave processes

has been solved by executing three different algorithms: the monoprocessor version, the parallel version without data reconfiguration and the parallel version using the data reconfiguration system. Comparing these three implemented algorithms, the mean time spent in executing 100 times the algorithm, varying the dimension of the grid, is (see Table 1).

The computing resources used for this project were those of the HLRS and LSS centers.

- HLRS Center:
  Processors: 400 Intel Xeon EM64T CPU (3.2GHz)
  Memory: 160 nodes * 1 GB + 40 nodes * 2 GB
  Network: Infiniband 1000 MB/s
- LSS Chair:
  Processors: 9 dual- and 8 quad- core AMD Opteron (2.2 GHz)
  Memory: 9 nodes * 4 GB + 8 nodes * 16 GB
  Network: GBit-interface and Infiniband

In both cases, at most, 32 nodes (slaves processes) were used. The maximum matrix dimension of the 2D grid that defines the problem structure, was about 256×256 of long values although the system can work with any size and any dimension.

From the point of view of the final data topology, the best solution is to use the DFS method as grouping algorithm. If we compare both methods (as in Fig. 16), we can observe that the DFS method generates more compacts groups, while the BFS generates groups that are, theoretically, more scatter. Because of this reason, the execution time using the BFS algorithm is slightly greater.

Table 1 Execution time, in seconds, of the three version of the same algorithm. If the dimension of the problem increases (value D) in a factor of two, the monoprocessor version increases its execution time in a factor of 4, since the parallel version increases in a value of 2. The grouping algorithm used in this execution was the DFS. The best improvement achieved with this example was nearly 13%

Algorithm	$D = 8$	$D = 16$	$D = 32$	$D = 64$	$D = 128$
Monoprocessor	0.15	0.62	2.48	9.46	33.13
Without Reconfiguration	1.41	3.12	6.28	13.02	25.17
With Reconfiguration	1.29	2.85	5.42	11.46	23.29

Acknowledgements This work was carried out under the HPC-EUROPA++ project number 211437, with the support of the European Community—Research Infrastructure Action of the FP7 "Coordination and support action" Programme.

The author would also gratefully acknowledge the support during the visit given by the Department of Computer Science 10, Chair for System Simulation (LSS), University of Erlangen-Nuernberg.

References

1. Sander, L., Meakin, P., Ramanlal, P., Ball, R.C.: Ballistic deposition on surfaces. Phys. Rev. A **1**, 34 (1986)
2. Krug, J.: The columnar growth angle in obliquely evaporated thin films. Matterwissen Werkstofft. **26**, 22–26 (1995)
3. Vicsek, T.: Fractal Growth Phenomena. World Scientific, Singapore (1989)
4. Hara, T.: A stochastic model and the moment dynamics of the growth of amorphous films. J. Theor. Biol. **1**, 109 (1984)
5. Meakin, P., Krug, J., Kassner, K., Family, F.: Laplacian needle growth. Europhys. Lett. **1**, 27 (1993)
6. Krug, J.: Origins if Scale Invariance in Growth Processes. Taylor & Francis, London (1997)
7. Oliva, A.I., Anguiano, E., Sacedón, J.L., Aguilar, M., Méndez, J.A., Aznérez, J.A.: Extended statistical analysis of rough growth fronts in gold films prepared by thermal evaporation. Phys. Rev. B **60**, 2720–2727 (1999)
8. Mandelbrot, B.B.: The Fractal Geometry of Nature. Freeman, New York (1982)
9. Meakin, P.: Fractals, Scaling & Growth Far from Equilibrium. Cambridge University Press, Cambridge (1998)
10. Family, F., Vicsek, T.: Dynamics of Fractal Surfaces. World Scientific, Singapore (1989)
11. Mullins, W.W.: Solid Surface Morphologies Governed by Capillarity, Metal Surfaces Structure, Energetics and Kinetics. Gjostein, N.A., Robertson, W.D. (1963)
12. Michely, T., Krug, J.: Islands, Mounds, and Atoms. Patterns and Processes in Crystal Growth Far from Equilibrium. Springer Series in Surface Sciences, vol. 42. Springer, Berlin (2004)
13. Castañeda-Marroquín, C., Navarrete, C.B., Ortega, A., Alfonseca, M., Anguiano, E.: Parallel Metropolis-Montecarlo simulation for Potts model using an adaptable network topology based on dynamic graph partitioning. In: ISPDC '08 (2008)
14. Andrew, R.S.: Reliability in LAM/MPI
15. Burns, G., Daoud, R., Vaigl, J.: LAM: An open cluster environment for MPI
16. Forum, M.P.I.: MPI: A message-passing interface standard. Technical Report UT-CS-94-230 (1994)
17. Mascagni, M., Chi, H.: Parallel linear congruential generators with Sophie-Germain moduli. Parallel Comput. **30**(11), 1217–1231 (2004)

18. Culler, D.E., Gupta, A., Singh, J.P.: Parallel Computer Architecture: A Hardware/Software Approach. Kaufmann, San Francisco (1997)
19. Alexander, A., Mark, C., Rejniak, K.A.: Single-Cell-Based Models in Biology and Medicine. Birkhäuser, Basel (2007)
20. Graner, J.G.F.: Simulation of biological cell sorting using a 2-dimensional extended Potts model. Phys. Rev. Lett. **69**(13), 2013–2016 (1992)
21. Poplawski, N.J., Swat, M., Gens, J.S., Glazier, J.A.: Adhesion between cells, diffusion of growth factors, and elasticity of the aer produce the paddle shape of the chick limb. Phys. A, Stat. Theor. Phys. **373**, 521–532 (2007)
22. Turner, S., Sherratt, J.A.: Intercellular adhesion and cancer invasion: a discrete simulation using the extended Potts model
23. Rosenbluth, M., Teller, A., Teller, E., Metropolis, N., Rosenbluth, A.: Equation of state calculations by fast computing machines. J. Chem. Phys. **21**(13), 1087–1092 (1953)
24. Chen, N., Glazier, J.A., Izaguirre, J.A., Alber, M.S.: A parallel implementation of the cellular Potts model for simulation of cell-based morphogenesis. Comput. Phys. Commun. **176**(11–12), 670–681 (2007)
25. Cercato, F.P., Mombach, J.C.M., Cavalheiro, G.G.H.: High performance simulations of the cellular Potts model. In: High Performance Computing Systems and Applications, Annual International Symposium on, 0:28 (2006)
26. Gusatto, E., Mombach, J.C.M., Cercato, F.P., Cavalheiro, G.H.: An efficient parallel algorithm to evolve simulations of the cellular Potts model. Parallel Process. Lett. **15**(12), 199–208 (2005)
27. Chen, N., Glazier, J.A., Izaguirre, J.A., Alber, M.B.: A parallel implementation of the cellular Potts model for simulation of cell-based morphogenesis. Comput. Phys. Commun. **176**(11–12), 670–681 (2007)
28. Canal, R., Parcerisa, J., Gonzalez, A.: Dynamic cluster assignment mechanisms. In: HPCA, p. 133 (2000)
29. Bhargava, R., John, L.: Improving dynamic cluster assignment for clustered trace cache processors. Technical report (2003)
30. Amiri, K., Petrou, D., Ganger, G., Gibson, G.: Dynamic function placement in active storage clusters. Technical report (1999)
31. Lastovetsky, A.: Scientific programming for heterogene us systems—bridging the gap between algorithms and applications. In: PARELEC'06 IEEE Proceedings, pp. 3–8 (2006)
32. von Hanxleden, R., Scott, L.R.: Load balancing on message passing architectures. J. Parallel Distrib. Comput. **13**(3), 312–324 (1991)
33. Nicol, D.M., Saltz, J.H.: Dynamic remapping of parallel computations with varying resource demands. IEEE Trans. Comput. **37**(9), 1073–1087 (1988)
34. Kumar, S., Das, S.K., Biswas, R.: Graph partitioning for parallel applications in heterogeneous grid environments. In: IPDPS '02: Proceedings of the 16th International Symposium on Parallel and Distributed Processing, p. 66 (2002)
35. Bacigalupo, D.A., Jarvis, S.A., He, L., Spooner, D.P., Nudd, G.R.: Comparing layered queuing and historical performance models of a distributed enterprise application. In: IASTED International Conference on Parallel and Distributed Computing and Networks, pp. 608–613 (2005)
36. Bacigalupo, D.A., Jarvis, S.A., He, L., Spooner, D., Pelych, D., Nudd, G.R.: A comparative evaluation of two techniques for predicting the performance of dynamic enterprise systems. In: PARCO, pp. 163–170 (2005)

IV
Computational Fluid Dynamics

Direct Numerical Simulations of Turbulent Shear Flows

Richard D. Sandberg

Abstract This contribution is concerned with the simulation of turbulent shear flows. First, direct numerical simulations (DNS) are introduced, focusing on the challenges associated with computing turbulent flows without any empirical modelling. The numerical method of one of our DNS codes is presented in detail and its performance on different HPC systems is discussed, in particular issues arising with multi-core architectures. An overview of DNS-based research conducted in the Aerodynamics and Flight Mechanics Research group at the University of Southampton is then given, in an attempt to highlight how large-scale DNS performed on HPC resources can be used as a 'virtual wind tunnel', i.e. to conduct numerical experiments that cannot be performed in traditional laboratories. A range of different physical problems is discussed, including the noise generation of jets and turbulent flows over airfoils, turbulent structure generation in mixing layers and supersonic wakes, the sensitivity of turbulent spots to Mach number and wall temperature, turbulent breakdown of vortex rings, the influence of background turbulence on far-field wakes, and the evolution of wingtip vortices.

1 Introduction

Direct numerical simulation (DNS) denotes the solution of the fully nonlinear, time-dependent Navier–Stokes equations without any empirical closure assumptions. Because no modeling assumptions are made, DNS must resolve all length and time scales present in the physical problem being considered. This is in contrast to traditional computational fluid dynamics (CFD) approaches, where turbulence is modelled using Reynolds-averaged Navier–Stokes (RANS) closures, resulting in time averaged flow fields, and to large-eddy simulations (LES) which depend on sub-grid

Richard D. Sandberg
Aerodynamics and Flight Mechanics Research Group, School of Engineering Sciences, University of Southampton, Southampton, SO17 1BJ, UK, e-mail: sandberg@soton.ac.uk

M. Resch et al. (eds.), *High Performance Computing on Vector Systems 2009*,
DOI 10.1007/978-3-642-03913-3_12

scale models to account for the smallest scales that cannot be resolved directly by the simulation. Modelling uncertainties from RANS and LES are avoided in DNS and numerical experiments based on first principles can be conducted. However, this comes at the price of high computational cost, which directly scales with the Reynolds number, an important non-dimensional parameter defined as $Re = UL/\nu$, where U and L are characteristic velocity and length scales, respectively, and ν denotes the kinematic viscosity. Turbulence is a nonlinear process that features a wide range of time and length scales. The largest scales in turbulence are mostly determined by the geometry of the physical problem under consideration, and therefore do not depend on the Reynolds number. However, for the case of homogeneous isotropic turbulence, the smallest scales of turbulence η_K are considered universal and isotropic, and can be shown on grounds of dimensional analysis [1] to be related to the Reynolds number as

$$\frac{\eta_K}{L} = Re^{-\frac{3}{4}}. \tag{1}$$

For isotropic three-dimensional turbulence, the number of grid points $N \propto L/\eta_K$ that is required for a given simulation is thus proportional to $Re^{9/4}$. Further factoring in the change in timestep required to resolve the Kolmogorov timescale for increasing Reynolds number, the total computational effort actually is proportional to Re^3.

The implication of this Reynolds number scaling is, when including the additional cost of an efficient algorithm, that for each doubling of Reynolds number from what is currently feasible, roughly 11 times more numerical operations are required. If we can assume that available computer power keeps doubling every 18 month, the Reynolds number can thus be doubled every 5–6 years. This in term means that, even on the latest high-performance computing (HPC) systems, we are still far away from being able to perform DNS for Reynolds numbers typical for aerospace applications. Hence, despite the dramatic increase of computing power over the last decades, DNS has mostly been used for computing canonical flows at moderate Reynolds number, rather than being used as an engineering design tool. Nevertheless, for flows that could be reliably computed, DNS performed on HPC systems have provided a wealth of data, not obtainable by traditional laboratory experiments, and have helped gaining insight into fundamental mechanisms in fluid mechanics [2] and, more recently, aeroacoustics [3]. DNS data have also been heavily used for developing, validating, or improving turbulence models [4].

Following its early successes, DNS has now reached the stage where it no longer is employed to reproduce experimental results. DNS is now frequently used as what might be called a *Virtual Wind Tunnel*, which can be defined as a numerical experiment that cannot be performed in the laboratory, for the purpose of gaining insight into the fundamental physics of turbulence or noise generation. These type of Virtual Wind Tunnel simulations can involve either choosing flow geometries or boundary conditions that cannot be realized in practice. Examples of Virtual Wind Tunnel simulations are discussed in Sect. 5.

2 Numerical Method

Because DNS is required to resolve the entire range of spatial and temporal scales in turbulence, the numerical algorithms employed need to satisfy stringent requirements which include (i) accuracy and consistency of the numerical method (i.e. a small truncation error for a given grid size and time step is required and the discretized equations need to approximate the original PDEs), (ii) waves need to be resolved with minimal amplitude and phase errors (dissipation and dispersion properties of the scheme), (iii) the scheme must be stable, and (iv) the scheme must be efficient (i.e. high ratio of accuracy to computational cost). Thus, DNS hardly ever uses commercial CFD software; DNS are rather performed with purpose-written research codes, highly optimised for certain flow-types and HPC architectures.

The numerical method of a recently developed research code, representative of some of the codes that have been used for DNS of compressible flows in Southampton, is presented in the following. The High-Performance Solver for Turbulence and Aeroacoustics Research (HiPSTAR) is a multi-block code that solves the full non-linear compressible Navier–Stokes equations in curvilinear cylindrical co-ordinates and currently features a 4th-order accurate central finite difference scheme with a 4th-order accurate Carpenter [5] boundary scheme, ultra low-storage 4th-order Runge–Kutta time-integration [6], skew-symmetric splitting of the non linear terms [7], and non-reflecting boundary conditions [8]. In the azimuthal direction, a spectral method can be employed, using the FFTW library or Schwarztrauber-based FFTs, enabling a state-of-the-art axis treatment that exploits parity conditions of individual Fourier modes. In order to take full advantage of massively parallel HPC systems, a domain decomposition approach was chosen and the code was parallelized using MPI. Furthermore, the algorithm was designed such that the memory requirement of the code was minimized.

3 Performance on Distributed Memory Systems

Within the framework of the HPC-Europa++ transnational access programme, HiPSTAR was ported to several computing platforms to conduct performance studies. Initially, only distributed memory systems, i.e. clusters, were considered. Performance was tested on Cacau at HLRS, Stuttgart, a NEC Intel Xeon cluster with dual-core processors, the IC1 in Karlsruhe with two Intel Xeon quad-core CPUs per node, and HECToR, a Cray XT4 at EPCC with four dual-core AMD Opteron processors per blade. First, a test case was selected with roughly 3.5×10^6 grid points, i.e. a relatively small simulation that was designed to benefit of cache effects at fairly small CPU numbers. The performance of the code on all clusters used is shown in Fig. 1. Considering the performance on the clusters with dual core CPUs first (Fig. 1, left), it is evident that the code does not scale linearly when increasing the core number from one to two. However, from that point on, the code appears to scale close to linearly up to 1,024 cores, as shown by using the simulation on two cores as reference (dashed lines). When requesting one core per CPU on HECToR, scaling is excellent over the entire range of cores. Looking at the performance on the IC1 cluster, it appears that the inter-node performance is poor, in particular when considering the region between two and eight cores. However, for larger number of cores, performance is very good and super-linear speed-up is observed. By requesting less cores per node, it is demonstrated that the MPI parallelization is highly efficient, i.e. linear scaling is achieved for the entire range when one core per node is used.

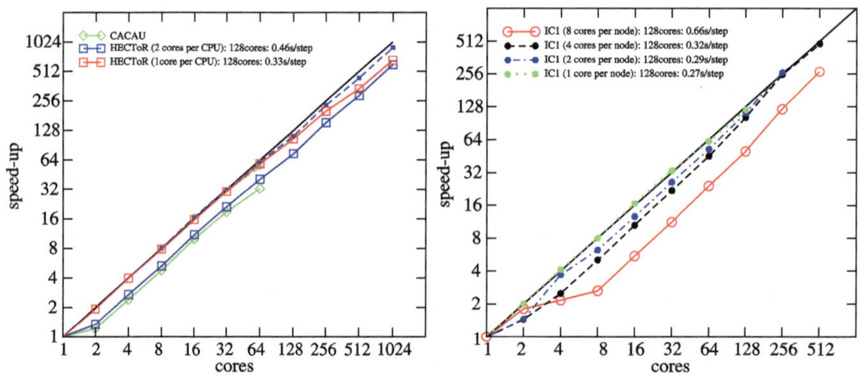

Fig. 1 Speed-up of the code on distributed memory systems for a test case with 3.5×10^6 grid points; Cacau and HECToR (left), IC1 (right)

Subsequently, a much larger test case was selected with 100×10^6 points, which is typical for future production runs. For this case, the simulations could not be run on less than 32 cores, hence internodal issues could not be tested. As can be seen from Fig. 2 (left), the performance of the code is excellent on all platforms tested. The final test case was set up such that a constant number of 480,000 grid points (i.e. equal memory size) per core was maintained. Note that at 1024 cores, this resulted in a simulation with 0.49×10^9 points. Using this approach, cache effects could be eliminated and the parallelization could be tested more rigorously. Figure 2 (right) reinforces the above findings, showing that a nearly constant performance could be achieved by selecting only one core per CPU or node, while performance decreased the more cores per CPU or node were chosen.

Overall, the scaling of the code on distributed memory systems was satisfactory, albeit the internodal performance loss is an issue that will require further attention.

4 Performance on a Vector System

For efficient use of the NEC SX-8, a HPC system with vector architecture, several modification to the code were required. The code was profiled using ftrace in order

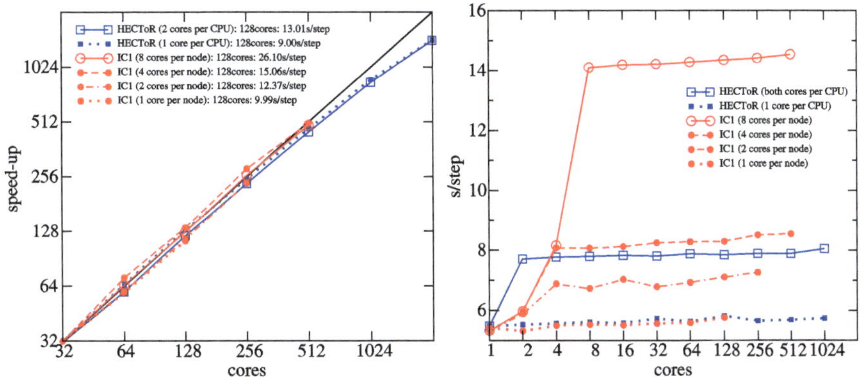

Fig. 2 Performance of the code on distributed memory clusters; speed-up for a large test case with 100×10^6 grid points (left); time per step using a constant number of 480,000 grid points per core (right)

to determine possible bottlenecks, with a main focus on vectorization of individual FORTRAN do-loops, vector length and the vector operation ratios within individual routines. Several FORTRAN loops were modified and resulted in significant performance gains. Furthermore, the FFTW routines, the best choice for the scalar systems, were replaced by vectorized FFT routines based on those by Schwarztrauber. With all modifications, the performance of the code on the NEC SX-8 was up to 13 times better than on any scalar system tested, achieving 4870 MFLOPs per process for a test case with 20×10^6 grid points.

5 The 'Virtual Wind Tunnel'

In this section, an overview of DNS-based research conducted in the Aerodynamics and Flight Mechanics Research group at the University of Southampton is given, to showcase how large-scale DNS performed on HPC resources can be used as a virtual wind tunnel. For the examples chosen, the physically unattainable features are not due to modifications of the governing equations, but to the flow configuration and boundary conditions.

5.1 Supersonic Axisymmetric Wakes

DNS have been performed of supersonic axisymmetric wakes with the main goal of gaining an understanding of the physical mechanisms responsible for the formation of energetic large-scale coherent structures. These structures have been observed in specifically designed laboratory experiments [9], however, their origin and evolution are not yet fully understood. A deeper insight in the relevant aerodynamic mechanisms in supersonic wakes are important for drag reduction of supersonic vehicles, and for manoeuvrability using flow control. HiPSTAR was specifically written for these kinds of flows and was therefore used for the simulations of a wake at Mach number of $M = 2.46$ and a Reynolds number, based on the freestream velocity and the diameter of the cylinder, $Re_D = 100{,}000$. This case was also chosen to enable validation of the novel code, as comparison with previous data [10] was possible. The simulation was discretized with 1,200 and 160 points in the streamwise and

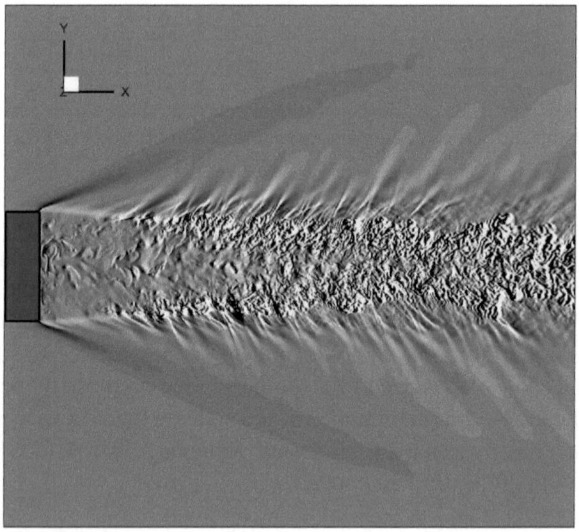

Fig. 3 Contours of instantaneous density gradients for a wake downstream of a cylinder; $Re_D = 100{,}000$, $M = 2.46$

radial directions, respectively, and 128 full Fourier modes in the azimuthal direction, resulting in approximately 50×10^6 collocation points. For a qualitative view of the flow field, instantaneous contours of the density gradient after 100,000 timesteps are shown in Fig. 3. A fast laminar-turbulent transition within the initial shear layer can be observed, producing shocklets in the outer shear layers and a turbulent developing wake downstream of the mean reattachment region. Animations of the initial start-up were also used to obtain an indication of the dominant mechanisms responsible for structure generation. The data compare well with the reference data, in spite of utilizing full Fourier transforms for the discretization of the azimuthal direction in the current case, versus assuming symmetric Fourier transforms (i.e. only using the sine or cosine transforms, depending on the symmetry of a variable), as in the previous study [10]. This suggests that employing the computationally less expensive symmetric Fourier transforms is a good approximation. Further simulations are planned at $Re_D = 200{,}000$ because linear stability analysis indicated that additional instability mechanisms might exist at the higher Reynolds number. These simulations will require a significantly larger number of points because a turbulent inflow will be required, and will only be possible on current HPC systems.

5.2 Turbulent Flow over Airfoil Trailing Edges

DNS of turbulent flow over airfoil trailing edges were conducted with the aim of understanding trailing-edge (TE) noise and airfoil self-noise generation, i.e. noise produced by the interaction of the airfoil with its own boundary layers and wakes. Trailing-edge noise can be shown to be one of the dominant contributors of the overall noise produced by airfoils. Insights gained from this research are expected to help design quieter aircraft, propulsion systems and wind turbines. The DNS were conducted with a code similarly structured to HiPSTAR, although a finite-difference scheme is used for the spanwise direction.

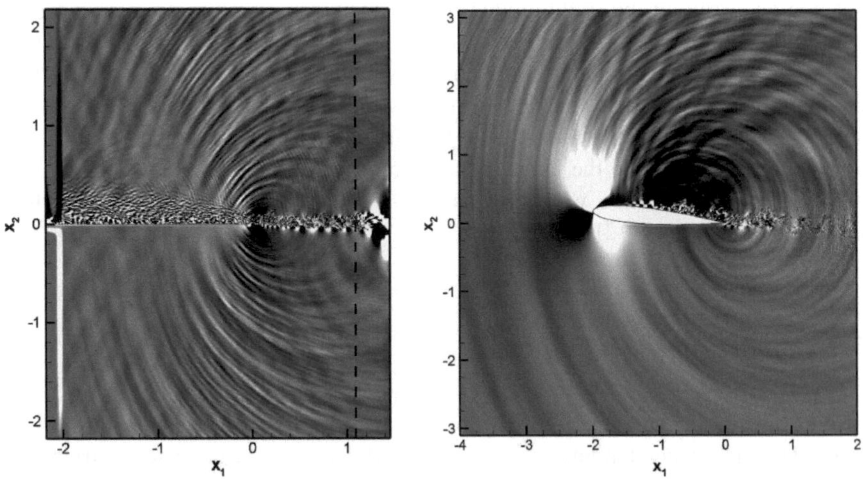

Fig. 4 Instantaneous contours of dilatation for turbulent flow past flat plate (left) with levels $[2 \times 10^{-3}; 2 \times 10^{-3}]$, and flow past a NACA-0012 airfoil at $\alpha = 5°$ (right) with levels $[3 \times 10^{-2}; 3 \times 10^{-2}]$

DNS were initially conducted of turbulent flow over flat plates, allowing for the data to be used to evaluate key elements of classical trailing-edge noise theories derived for that particular flow configuration. Figure 4(left) shows instantaneous contours of dilatation rate from a DNS of compressible turbulent flow at $M = 0.6$ with a maximum $Re_{\delta^*} = 2{,}200$ at the trailing edge. 2,001 and 609 non-equidistantly spaced points were used in the x_1- and x_2-directions, respectively, with the finest resolution at the TE. In the spanwise direction, 87 equidistantly spaced points were used. This resulted in a total of 106×10^6 gridpoints and required 80,000 CPU hours on HPCx, an IBM p5-575, using 512 processors. The unique feature of this DNS is that a turbulent boundary layer was only generated on the top surface, while a laminar boundary layer was specified for the bottom side. The consequence of this set-up, which most likely cannot be reproduced in a laboratory experiment, is that the trailing edge noise can be separated from other noise sources. On the bottom side, only the acoustic waves generated at the TE are present and can therefore be observed clearly in Fig. 4(left). On the top surface, the turbulent boundary layer itself appears to be a source of noise, with the resulting acoustic waves interacting with the trailing edge noise. The main outcomes of this simulation were a modification to the analytical description of the skinfriction in the vicinity of the TE, the conclusion that a simplified two-dimensional theory could be used to predict noise generation accurately for low-frequencies, and the validation of some of the individual assumptions made in the derivation of classical TE noise theories [11].

The next step was to conduct DNS of flow past a NACA-0012 airfoil at $\alpha = 5°$ for $Re_C = 50{,}000$ and $M = 0.4$, using 2570×692 non-equidistantly spaced grid points in the lateral and normal directions, respectively, with a greater number of tangential points placed on the suction side of the airfoil and the smallest grid spacing occurring at the trailing edge. In the spanwise direction, 96 equidistantly spaced points were used, resulting in a total of approximately 170×10^6 points. Figure 4(right) shows instantaneous contours of dilatation rate for the full airfoil configuration, illustrating one of the main conclusions of the study: the radiated sound field is highly asymmetric about the airfoil chord due to the presence of additional noise sources in the laminar-turbulent transition/reattachment region. The data from this simulation were also used to evaluate elements of existing airfoil noise prediction tools [12].

5.3 Compressible Mixing Layer

Mixing layers are canonical flows representative of jets, wakes and other shear flows. Detailed understanding of the origins and dynamics of large-scale structures in mixing layers is essential for jet-noise reduction, improved mixing for combustion, and many other applications. Landmark experiments of Brown and Roshko [13] illustrated the development of large rollers with high spanwise coherence. Despite significant subsequent research, the origin of the striking two-dimensionality of the structures observed in the experiments is still not clear. Most previous numerical studies have lacked the splitter plate separating the two streams prior to mixing due to limitations of computing resources, thereby omitting a physical feature possibly responsible for the presence of spanwise-coherent structures. A DNS was therefore conducted of a mixing layer driven by two streams with a velocity ratio of ten, which included the splitter plate to assess whether the plate might indeed be instrumental in creating large-scale spanwise-coherent rollers [14]. A turbulent boundary layer was present on the side of the fast stream at $M = 0.6$, while the slow stream was laminar. The same research code as for the trailing-edge simulations of Sect. 5.2 was used. The computational grid comprised $4{,}001 \times 513 \times 225$ points in the streamwise, normal and spanwise directions, respectively, giving a total of 462×10^6 grid points and was run on 2,048 cores on HECToR for 72,000 timesteps, making it one of the largest compressible CFD simulations performed in the UK to date. Instantaneous contours of streamwise density gradients for two time instants,

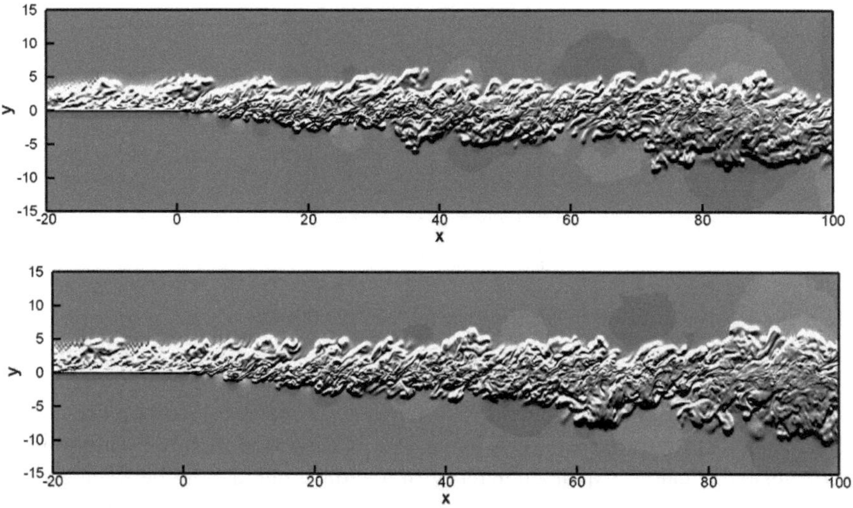

Fig. 5 Instantaneous contours of streamwise density gradients for two instants in time

shown in Fig. 5, do not display the spanwise-coherent structures found in the original experiments. This suggests that the presence of a splitter plate is not sufficient by itself to generate these structures in the early development of the mixing layer. In the current case, an infinitely thin plate was used. Thus, to entirely rule out the splitter plate as cause for Brown–Roshko structures, a further simulation with a finite thickness plate will need to be computed.

5.4 Jet Noise

A large body of research exists for jet noise, however, there remain open questions, including a detailed description of the noise generation mechanisms in subsonic jets, and the influence of upstream disturbances on the noise generation. Large-scale DNS were therefore performed of round jets to contribute to the understanding of noise generation in jets in order to ultimately achieve a reduction of jet-noise. The DNS were conducted with a modified version of the code used for the flows described in 5.2 and 5.3. The largest simulation required $561 \times 561 \times 1350$ grid points, giving 425×10^6 total points, and was run on HPCx. An example of the sound field radiated from the subsonic jet is shown in Fig. 6 by instantaneous contours of the dilatation rate. In this case the flow at the inflow boundary was perturbed by a broad

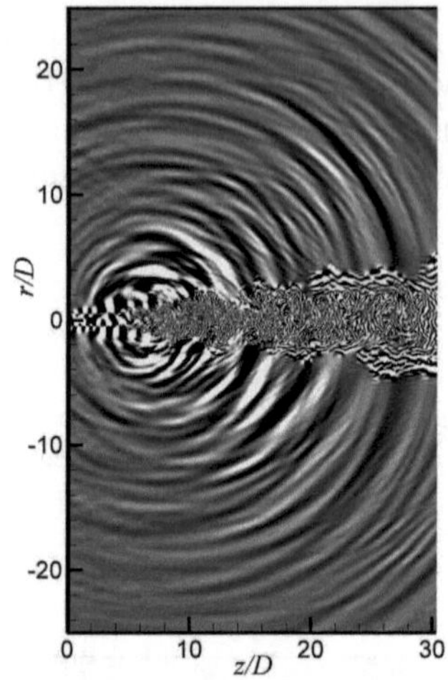

Fig. 6 Instantaneous contours of dilatation rate for round jet simulation

range of modes. The results obtained so far indicate that the low-frequency noise at the angles close to the jet downstream axis is generated mainly by a nonlinear mechanism that can be explained in terms of nonlinear interactions between various primary instability waves. High-frequency noise, on the other hand, seems to originate from the breakdown of the potential core into smaller structures with ultimate transition to turbulence [15]. To further investigate the influence of disturbance amplitudes and possibly of fully turbulent inflow, a number of very large DNS will still need to be performed.

5.5 Turbulent Spots in Supersonic Boundary Layers

For hypersonic vehicles, accurate prediction of the laminar-turbulent transition in a supersonic or hypersonic flow is essential for quantifying heat transfer and drag. These quantities need to be known in order to decide on suitable materials enabling sustained hypersonic flight. Turbulent spots play an important role in transition as they are one route by which a strongly perturbed boundary layer finally becomes turbulent.

Using the Navier–Stokes solver also used for the previously described cases, extended with a TVD-scheme to account for shocks occurring at supersonic speeds, several DNS were conducted to investigate the influence of the Mach number and the wall-boundary condition on the development of turbulent spots [16]. Computational grids with up to 50×10^6 points were used and the simulations were run on HPCx. In Fig. 7, instantaneous iso-surfaces of $(\partial u_i/\partial x_j)(\partial u_j/\partial x_i)$ and $u' = u - u_{lam}$ are shown for four cases, $M = 3$ with an adiabatic (hot) wall, $M = 3$ with a cold wall ($T_w = T_\infty$), $M = 6$ with an adiabatic wall, and $M = 6$ with a cold

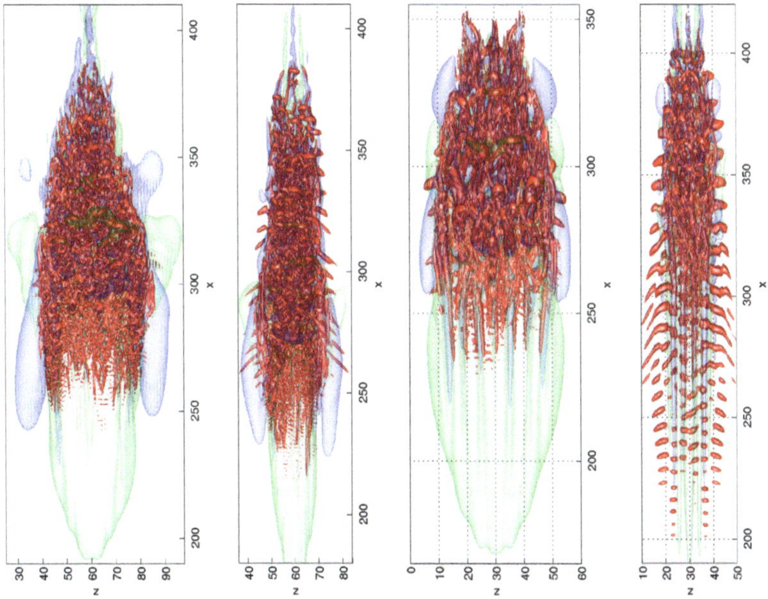

Fig. 7 Instantaneous iso-surfaces of $(\partial u_i/\partial x_j)(\partial u_j/\partial x_i) = -8\times 10^{-4}$ in red, and $u' = u - u_{lam} = \pm 0.02$ in light blue and green for $M = 3$ adiabatic; $M = 3$ cold; $M = 6$ adiabatic; $M = 6$ cold (left to right)

wall. It can clearly be discerned that the structure of the turbulent spots is strongly affected by Mach number and the wall boundary condition. The spreading angle of the turbulent spots gives an indication of the length of the transition process and is therefore an important factor. As seen in Fig. 7, changing from an adiabatic to a cold-wall boundary condition reduces the spreading angle by 20 to 30%.

5.6 Turbulent Breakdown of Vortex Rings

The interest in vortex rings stems from the fact that their growth, instabilities and breakdown embodies prototypical turbulent flows. In addition, vortex rings are ubiquitous in engineering and biological flows. Investigating the turbulent breakdown of vortex rings therefore promises to advance the general understanding of breakdown to turbulence. An incompressible in-house code was used to perform DNS of a single vortex ring undergoing transition from a laminar state to an early stage of full turbulence. The solver uses second-order finite differences in space on a conventional staggered grid, and the Adams–Bashforth algorithm in time, with continuity imposed by applying pressure-correction methods with a parallel multigrid Poisson solver. For the largest case computed on HPCx, a computational grid with 512^3 points was required, resulting in 134×10^6 grid points.

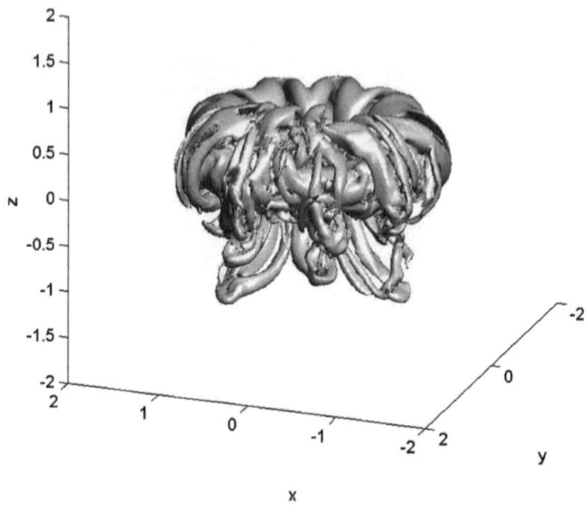

Fig. 8 Iso-surfaces of the second invariant of the velocity gradient tensor for a vortex ring

Figure 8 shows iso-surfaces of the second invariant of the velocity gradient tensor and highlights the secondary structures that have formed once the inner core instability waves reached sufficiently large amplitudes. The secondary structures form through tilting and stretching of the outer core vorticity and can be described as counterrotating neighbouring loops of vorticity. An important feature of these structures is that they cause localised detrainment of fluid [17].

5.7 Wing Tip Vortex Breakdown and Far Wakes

The generation of lift by fixed-wing aircraft gives rise to long-lived trailing vortices that may be hazardous to other aircraft. Current air traffic control rules are based on the observation that these vortices descend about 900 feet from their initial altitude, and then remain there; a flightpath that remains above that of the aircraft responsible for the wake is therefore believed to be safe.

Numerical simulations indicate that stable stratification has a strong effect on the motion of wake vortices in the atmosphere. 2-D studies found that for a large airliners landing in very stably stratified conditions, the wake vortices initially descend, but then rebound upwards. It is, however, possible that in practice, 3-D instabilities break up the vortices preventing this from happening. Therefore, 3-D simulations of a vortex pair perturbed with a long-wavelength Crow instability mode were performed.

These simulations were carried out by solving the incompressible Navier–Stokes equations in the Boussinesq approximation. This was done using a pseudospectral code, employing Fourier representation in space because periodic boundary conditions could be assumed in all directions. Computational grids with up to 260×10^6 collocation points were used for these simulations, which were carried out on the

HPCx and HECToR systems. For these cases, in the absence of any 3-D effects, rebound would take the vortices above their original height; however, Crow linking (as shown in Fig. 9) does indeed take place well before this can happen [18]. Nevertheless, it remains conceivable that atmospheric turbulence may not excite the Crow instability strongly enough for this to always be relied on in practice.

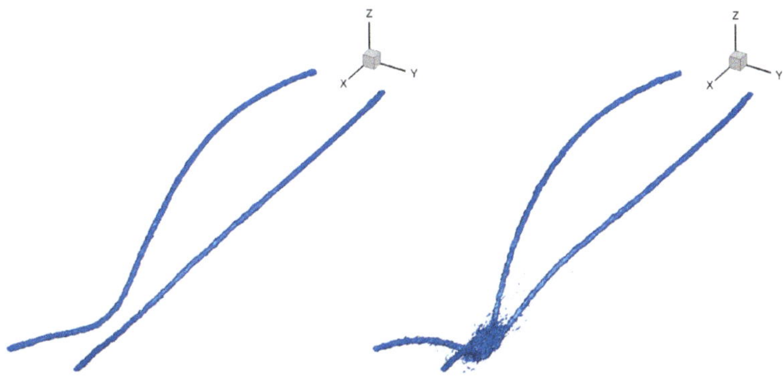

Fig. 9 Isosurfaces of vorticity magnitude for two instances in time

The same incompressible triple-spectral code was also used to perform DNS of time developing far axisymmetric wake at moderate Reynolds numbers, subjected to different isotropic turbulent backgrounds to study the influence of freestream turbulence on the wakes [19]. Because wakes of any three-dimensional object becomes axisymmetric sufficiently far downstream and might interact with some form of background turbulence, the fundamental study might contribute to a more general understanding of wake flows. The number of Fourier modes was $512 \times 2048 \times 512 \approx 537 \times 10^6$ points, with each time step taking approximately 50 seconds using 512 cores on HPCx. The results obtained so far indicate that the presence of freestream turbulence enhances the wake's decay rate, and that no new self-similar state is achieved due to background turbulence because the free-stream turbulence intensity and the wake's centreline velocity deficit decay at different rates.

6 Summary

A brief introduction of the challenges for direct numerical simulations of turbulent flows has been given. The numerical method of an in-house DNS code was described and its performance on various HPC architectures studied, revealing performance issues on multi-core architectures. DNS-based research within the Aerodynamics and Flight Mechanics Research group at the University of Southampton was presented. A range of different flows was discussed to showcase how DNS can be used as a virtual wind tunnel, and to demonstrate the perpetual need for state-of-the-art HPC resources.

Acknowledgements Code performance studies were supported by HPC Europa++ transnational access under project number 1264. The author is grateful to P.J. Archer, I.P. Castro, G.N. Coleman, R. Johnstone, L.E. Jones, J.A. Redford, E. Rind, N.D. Sandham, and V. Suponitsky for their contributions to this paper.

References

1. Kolmogorov, A.N.: The local structure of turbulence in incompressible viscous fluid for very large Reynolds numbers. Dokl. Akad. Nauk SSSR **40**(4) (1941)
2. Kim, J., Moin, P., Moser, R.D.: Turbulence statistics in fully developed channel flow at low Reynolds number. J. Fluid Mech. **177**, 133–166 (1987)
3. Freund, J.B.: Noise sources in a low-Reynolds-number turbulent jet at Mach 0.9. J. Fluid Mech. **438**, 277–305 (2001)
4. Spalart, P.R., Allmaras, S.R.: A one-equation turbulence model for aerodynamic flows. AIAA Paper 92–0439, January 1992. 30th AIAA Aerospace Sciences Meeting, Reno, USA
5. Carpenter, M.H., Nordström, J., Gottlieb, D.: A stable and conservative interface treatment of arbitrary spatial accuracy. J. Comput. Phys. **148**(2), 341–365 (1999)
6. Kennedy, C.A., Carpenter, M.H., Lewis, R.M.: Low-storage, explicit Runge–Kutta schemes for the compressible Navier–Stokes equations. Appl. Numer. Math. **35**, 177–219 (2000)
7. Kennedy, C.A., Gruber, A.: Reduced aliasing formulations of the convective terms within the Navier–Stokes equations for a compressible fluid. J. Comput. Phys. **227**, 1676–1700 (2008)
8. Sandberg, R.D., Sandham, N.D.: Nonreflecting zonal characteristic boundary condition for direct numerical simulation of aerodynamic sound. AIAA J. **44**(2), 402–405 (2006)
9. Bourdon, C.J., Dutton, J.C.: Planar visualizations of large-scale turbulent structures in axisymmetric supersonic separated flows. Phys. Fluids **11**, 201–213 (1998)
10. Sandberg, R.D., Hermann, F.F.: Numerical investigation of transitional supersonic axisymmetric wakes. J. Fluid Mech. **563**, 1–41 (2006)
11. Sandberg, R.D., Sandham, N.D.: Direct numerical simulation of turbulent flow past a trailing edge and the associated noise generation. J. Fluid Mech. **596**, 353–385 (2008)
12. Sandberg, R.D., Jones, L.E., Sandham, N.D.: Direct numerical simulations of noise generated by turbulent flow over airfoils. AIAA Paper 2008–2861, May 2008. 14th AIAA/CEAS Aeroacoustics Conference (29th AIAA Aeroacoustics Conference), Vancouver, Canada
13. Brown, G.L., Roshko, A.: On density effects and large structures in turbulent mixing layers. J. Fluid Mech. **64**, 775–781 (1974)
14. Sandham, N.D., Sandberg, R.D.: Direct numerical simulation of the early development of a turbulent mixing layer downstream of a splitter plate. J. Turbul. **10**(1), 1–17 (2009)
15. Suponitsky, V., Sandham, N.D.: Nonlinear mechanisms of sound radiation in a subsonic jet. AIAA Paper 2009–3317, May 2009. 15th AIAA/CEAS Aeroacoustics Conference (30th AIAA Aeroacoustics Conference), Miami, Florida
16. Redford, J.A., Sandham, N.D., Roberts, G.T.: Parametric study of compressible turbulent spots. In: ERCOFTAC Workshop: Direct and Large-Eddy Simulations 7 (DLES7), Trieste, Italy, September 8–10th 2008
17. Archer, P.J., Thomas, T.G., Coleman, G.N.: Direct numerical simulation of vortex ring evolution from the laminar to the early transition regime. J. Fluid Mech. **598**, 201–226 (2008)

18. Coleman, G.N., Johnstone, R., Yorke, C.P., Castro, I.P.: DNS of aircraft wake vortices: the effect of stable stratification on the development of the Crow instability. In: ERCOFTAC Workshop: Direct and Large-Eddy Simulations 7 (DLES7), Trieste, Italy, September 8–10th 2008
19. Rind, E., Redford, J.A., Castro, I.P.: Far axisymmetric wakes in turbulent streams. In: Sixth International Symposium on Shear Flow and Transport Phenomena, TSFP6, Seoul, Korea, June 2009

Large-Scale Flow Computation of Complex Geometries by Building-Cube Method

Daisuke Sasaki, Shun Takahashi, Takashi Ishida, Kazuhiro Nakahashi, Hiroaki Kobayashi, Koki Okabe, Youichi Shimomura, Takashi Soga, Akihiko Musa

Abstract Three-dimensional large scale incompressible flow simulation was conducted by Building-Cube Method (BCM) that is based on equally-spaced Cartesian mesh method. To exploit expected near-future high performance computer with massive number of processors, simple algorithms have been implemented in the BCM for mesh generation and flow solver. In this study, the capability of BCM for large-scale computation was demonstrated by solving formula 1 car model with around 200 million cells. The computation was conducted on Vector-Parallel Supercomputer system of NEC SX-9 at Cyberscience center of Tohoku University. The parallel efficiency of BCM with flat MPI and hybrid MPI on vector-parallel type system was also investigated.

1 Introduction

Nowadays, parallelization is one of key technologies to accelerate large-scale numerical computations. Although each high-performance computer system performance rapidly increases every year, the performance development has been accomplished as the increase of the number of processors in a system. Figure 1 shows

Daisuke Sasaki, Shun Takahashi, Takashi Ishida, Kazuhiro Nakahashi
Department of Aerospace Engineering, Tohoku University, Sendai 980-8579, Japan

Hiroaki Kobayashi, Koki Okabe
Cyberscience Center, Tohoku University, Sendai 980-8578, Japan

Youichi Shimomura
NEC Software Tohoku, Sendai 980-0811, Japan

Takashi Soga
NEC System Technology, Osaka 540-8551, Japan

Akihiko Musa
NEC Corporation, Tokyo 108-8001, Japan

M. Resch et al. (eds.), *High Performance Computing on Vector Systems 2009*,
DOI 10.1007/978-3-642-03913-3_13

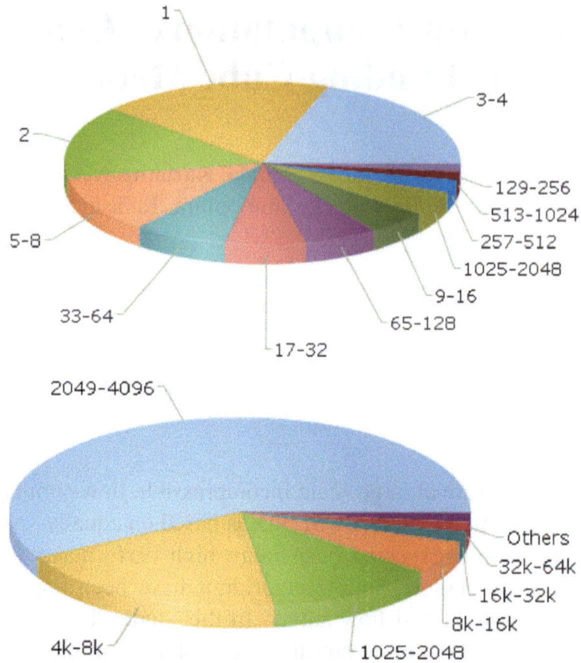

Fig. 1 Number of processors in a system of supercomputers [1] (above: 1993, below 2008)

number of processors in a system of supercomputers at 1993 and 2008 obtained by Top500 supercomputer sites [1]. As clearly shown in the figure, nowadays almost all the computers have more than thousands of processors, and this trend will be continued in the future. Therefore, parallel computational techniques for massive number of processors will be one of significant tasks to accelerate each application.

The Building-Cube Method (BCM) has been proposed to utilize massive number of processors effectively for various numerical simulations of fluids [2–5]. The approach is to divide entire flow domain into sub-domains named 'Cube'. Since each Cube is composed of the same number of equally-spaced Cartesian mesh, the computational load is equally distributed among Cubes. This feature realizes efficient and simple parallelization for mesh generation and flow simulation. The tactics of the BCM is to simplify the algorithms in all stages of flow computation (mesh generation, numerical simulation and post-processing) to conduct large-scale computations on near-future high-performance computers.

The capability of the BCM has been demonstrated in various 2D/3D compressible/incompressible flow problems. The developed mesh generation method significantly reduces the pre-processing time and enhances the reliability for complex geometry compared to unstructured mesh method we have developed. In this research, the capability of BCM for large-scale computation was demonstrated by solving flows around formula 1 car model with around 200 million cells on vector-super computer.

The large-scale computation of BCM is performed on vector-parallel supercomputer system of NEC SX-9 at Cyberscience Center of Tohoku University. The system has 16 nodes of SX-9 and each has 16 processing elements. The parallelization in each node can be parallelized by OpenMP but the parallelization over several nodes needs MPI parallelization. Therefore, two approaches of parallelization are compared in the study: flat-MPI and hybrid parallelizations. The former is entirely parallelized by MPI, and the latter is coupling between OpenMP and MPI, where OpenMP is used for processors in each node and MPI is used for nodes. The parallel efficiency of BCM with the two parallelization approaches will be also discussed.

2 Building-Cube Method

2.1 Overview

The Building-Cube Method (BCM) is based on equally spaced Cartesian mesh. A computational mesh of BCM consists of assemblage of various sized cuboids that are called 'Cube', and all cubes include completely the same number of equally-spaced Cartesian mesh called 'Cell' as shown in Fig. 2. In short, it is multi-block structure of cubes including cells that are equally-spaced Cartesian mesh. Vicinity of wall boundary is discretized by completely equally-spaced Cartesian mesh and the mesh resolution is coarsened with getting away from the wall boundary by changing 'Cube' size.

There are several advantages for equally-spaced Cartesian mesh compared to non equally-spaced Cartesian mesh. For instance, the solver can be simplified and the memory requirements can be minimized. To maintain its simplicity, BCM employs staircase representation of wall boundary. Instead of applying cut-cell approach and losing algorithm simplicity, very fine mesh is assumed to be used for wall boundary. This makes the algorithm simple as all cells can be simply classified into two groups that are wall and fluid.

2.2 Flow Solver

Governing equations are incompressible Navier-Stokes equations. Fractional-step method [6–8] is used with finite difference scheme on staggered arrangement. Time integration is performed by Adams-Bashforth second-order explicit method and Poisson equation of pressure is solved by successive over relaxation (SOR) method. Convective term is discretized by third-order upwind scheme [9] and other terms are discretized by second-order central difference scheme.

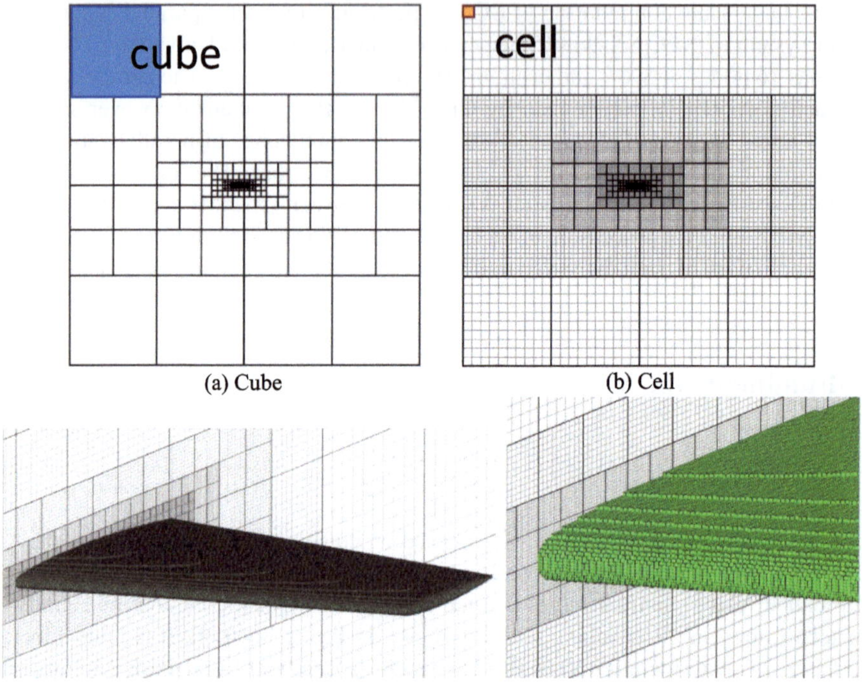

(a) Cube (b) Cell

(c) Cubes and cells around a wing (d) Staircase representation

Fig. 2 Cube and cells around a wing

Flow information needs to be exchanged between neighboring cubes because of the multi-block structure of BCM. First-order interpolation is performed at fringe cells between different-size cubes.

3 Code Optimization (Vectorization and Parallelization)

3.1 Vectorization

Vector-parallel supercomputer NEC SX-9 of 16 multi-nodes system, which was installed at Cyberscience center of Tohoku University in March 2008, is used for the present large scale parallel computation. Each node consists of 16 CPUs and each processor has over 100 Gflops of peak performance, 1 Tbyte shared main memory and 256 Gbyte per second memory bandwidth in a node. Ideal peak performance of a node is 1.6 Tflops. The data transfer rate between nodes archives 128 Gbyte per second and the ideal peak performance of the system is 26 Tflops.

To achieve high speed vector performance on vector machines, the flow solver needs to be highly tuned for vector operation. The linear solver for Poisson equa-

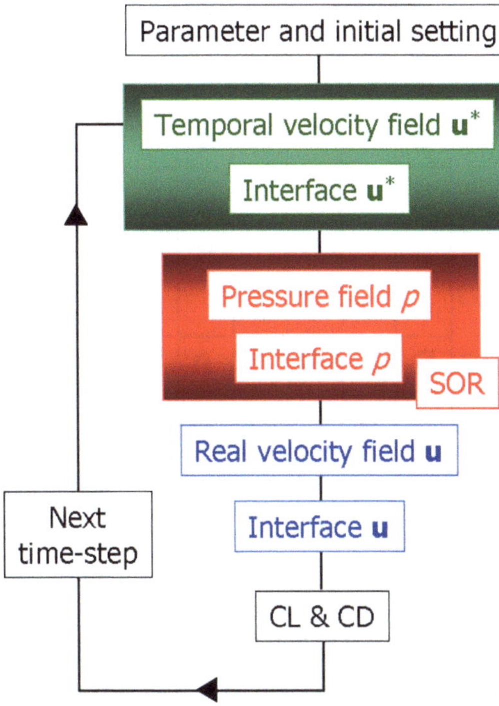

Fig. 3 Flowchart of BCM incompressible flow solver

tion tends to become a hotspot for the present MAC type algorithm. To accelerate this routine, SOR method used in the solver has to be optimized. Since data dependency in the iterative process of SOR routine prevents the vectorization, Red-Black coloring method is applied to SOR method to eliminate the data dependency. The schematic figure of Red-Black coloring is shown in Fig. 4. Almost all the loops of BCM consists of outermost loop of cube that is stored as unstructured data and inner three multiple loops of cells that are equally-spaced Cartesian meshes. Therefore, the present vectorization is applied to the inner three loops by loop extension technique of the conversion to one dimensional array. As a result of Red-Black coloring and other implementation, the linear solver in each cube achieved 99.95% of vectorization ratio.

3.2 Parallelization

Since SX-9 one node has 1 Tbyte large shared main memory, it is easy to perform relatively large-scale simulation with automatic parallelization or OpenMP parallelization. The advantage of OpenMP parallelization is no latency issues emerge during the parallel computation because of the thread parallel operation on shared

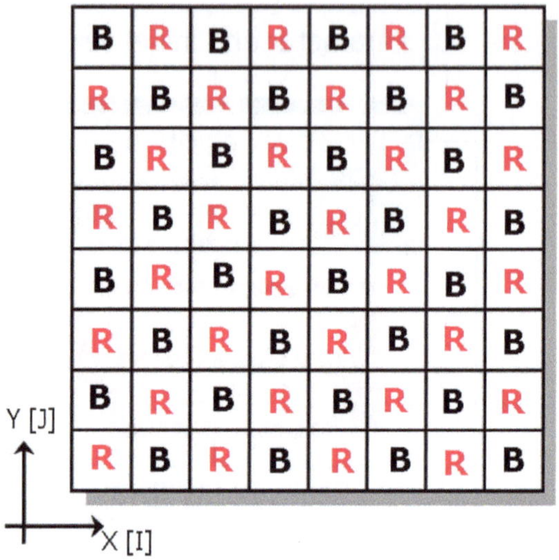

Fig. 4 Red-Black coloring in 8 × 8 Mesh in 2D

memory. However, in the case of multi-node parallel computation of SX-9, where the memory is distributed over nodes, MPI parallelization is essential to communicate with neighboring nodes.

When OpenMP parallelization is implemented for the present flow solver, the parallelization is applied to the outermost loop of cubes. Because BCM is multi-block structure of cubes in nature, most of flow computation is performed in each cube independently. Therefore, efficient parallelization can be realized easily when the parallelization is applied to the outermost loop of cubes. In the case of MPI parallelization, ideal efficient parallelization can be also realized when the number of cubes in each domain is same because of the same number of cells in each cube. In this study, the latest MPI 2.0 functions are used to achieve high speed communication: MPI_PUT and MPI_GET.

Although MPI is required for parallel computations over nodes, parallel computations inside each node can be performed by either OpenMP or MPI. Therefore, two methods of parallelization can be implemented: flat MPI and hybrid MPI. In the former parallelization, all the processors are parallelized by MPI. On the other hand, in the case of hybrid MPI, MPI parallelization is applied only between nodes and OpenMP parallelization is applied inside of a node as shown in Fig. 5.

4 Large-Scale Flow Computation

Large scale parallel flow simulation around a formula 1 racing car model is performed. The minimum spacing of the mesh is 7.32×10^{-4} based on overall length of the model and total number of cells is about 200 million. The computational cells

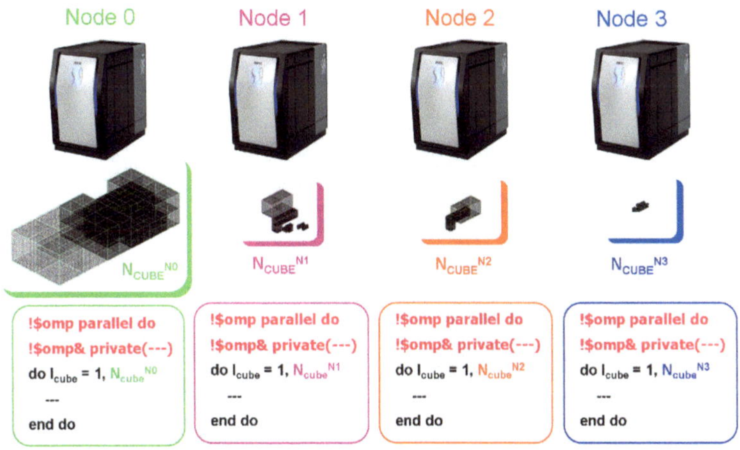

Fig. 5 Schematic of hybrid MPI approach for BCM

are visualized in Fig. 6. However, because of high-resolution computational BCM cells, coarsened cells are visualized in Figs. 6(a) and (b). The parallel computation is conducted by flat MPI and hybrid MPI to investigate the parallel efficiency for both the cases. Here, ground boundary is treated as non-slip wall and Reynolds number is set at 14.9 million.

The present result shows it can capture flow phenomena around a Formula 1 car such as flow acceleration over the rear wing and under the body, flow through the diffuser and roll up in the wake region as shown in Fig. 7. However, more precise computations will be needed for real flows by implementing the treatment of wheel rotation and moving ground boundary.

As for the parallel efficiency, OpenMP shows better performance than flat MPI for single node parallel computation because of no communication latency as shown in Fig. 8. On the other hand, flat MPI shows better performance than hybrid MPI for multi-nodes parallel computation as shown in Fig. 9. One of the possible reasons is that the amount of communication in flat MPI becomes smaller as the increase of the number of PEs, but communication time in hybrid MPI is almost constant regardless the number of PEs as indicated in Fig. 10. Thus, the scalability of flat MPI is better than hybrid MPI.

Parallel efficiencies using 8 and 16 nodes are examined by flat MPI parallelization as plotted in Fig. 11. The maximum performance of the present parallel computation on SX-9 is 130 times improvement in using 256 CPU. Since no consideration is made in the present domain decomposition, it generates many isolated regions on the domain assigned to each processor as shown in Fig. 12. This causes the random memory access when exchanging information with neighboring cubes, and it leads the deterioration of performance scalability. The parallel efficiency can be much improved by the introduction of rearrangement technique in domain decomposition.

Finally, the efficiency of vector-super computer of SX-9 is discussed. The SX-9 one node is composed of 16 processors, and its peak performance is 1600 Gflops.

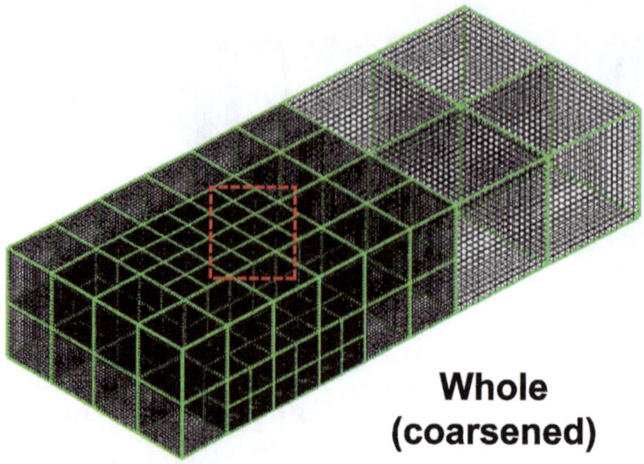

(a) Whole computational domain (coarsened for visualization)

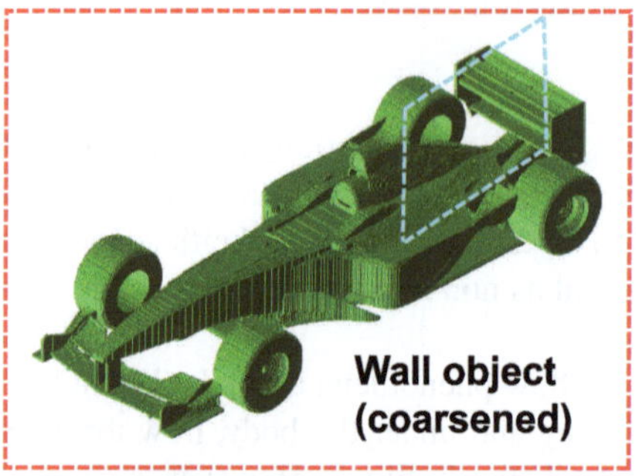

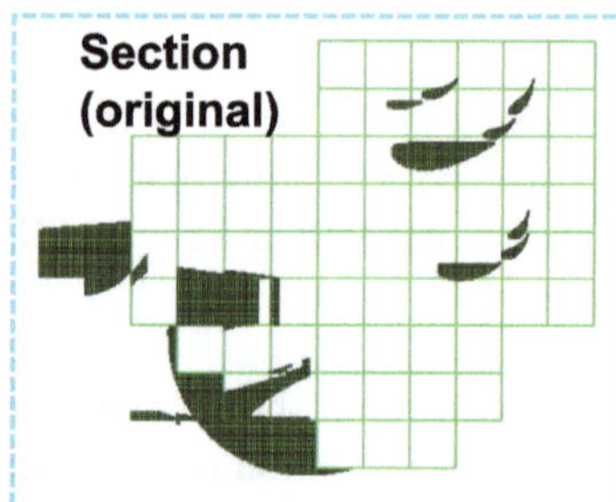

(b) Wall cells (coarsened for visualization) (c) Wall cells around rear wings

Fig. 6 Computational mesh around formula 1 racing car model

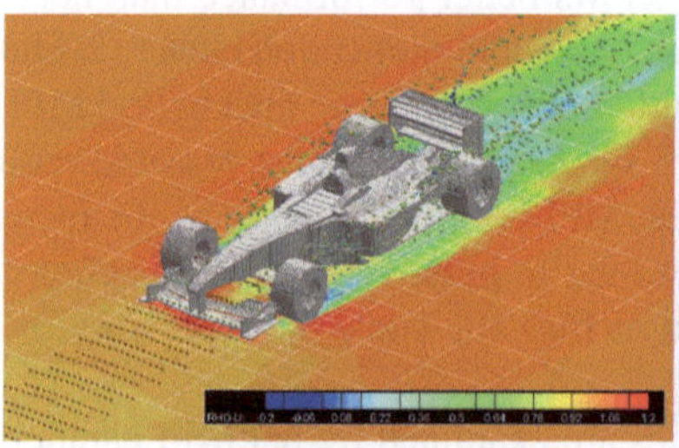

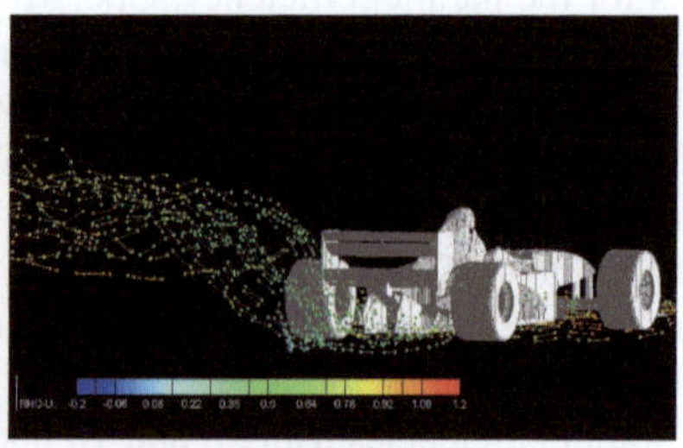

(a) Overview (b) Rear part

Fig. 7 Streamlines (time-averaged solution)

The comparison is made with scalar-parallel computer of NEC TX-7, which is composed of 64 cores inside of a node and its peak performance reaches 400 Gflops. CPU time for the present computation is compared as appeared in Fig. 13. The difference of the peak performance of the two systems is 4 times, but the actual time difference is nearly 8 times. This is due to the large advantage of vector operations

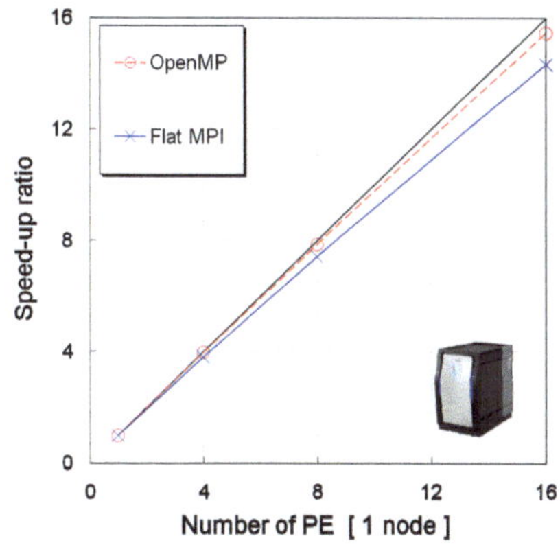

Fig. 8 Speed-up ratios of flat MPI and OpenMP on 1 node

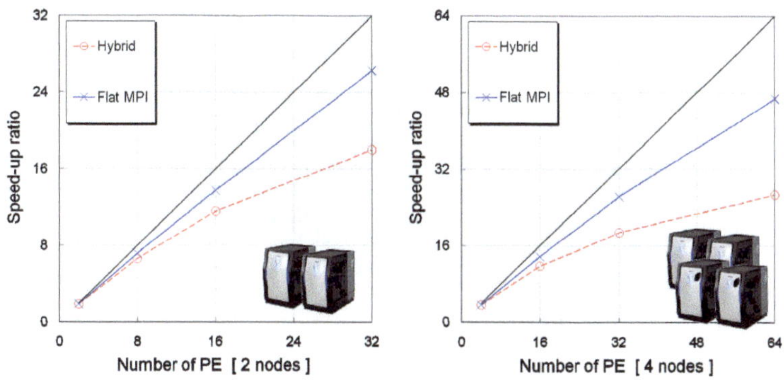

Fig. 9 Speed-up ratios of flat MPI and hybrid MPI (left: 2 nodes, right: 4 nodes)

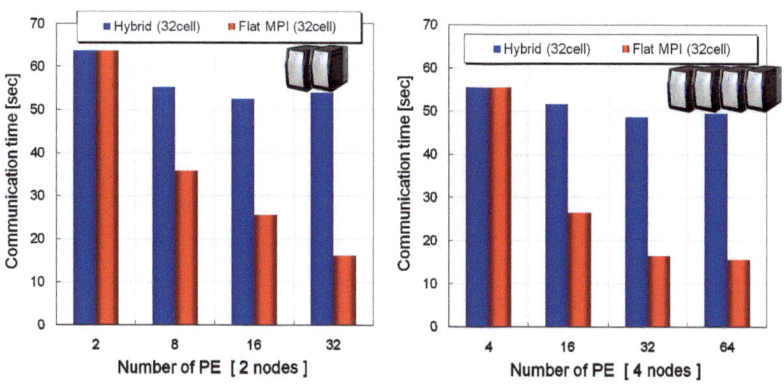

Fig. 10 Communication time of flat MPI and hybrid MPI (left: 2 nodes, right: 4 nodes)

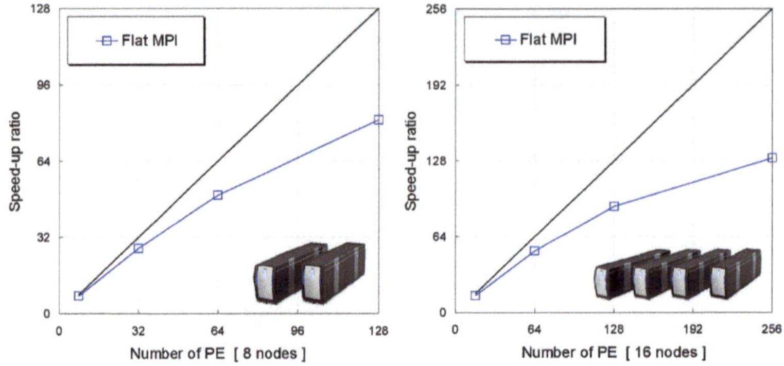

Fig. 11 Speed-up ratios of flat MPI (left: 8 nodes, right: 16 nodes)

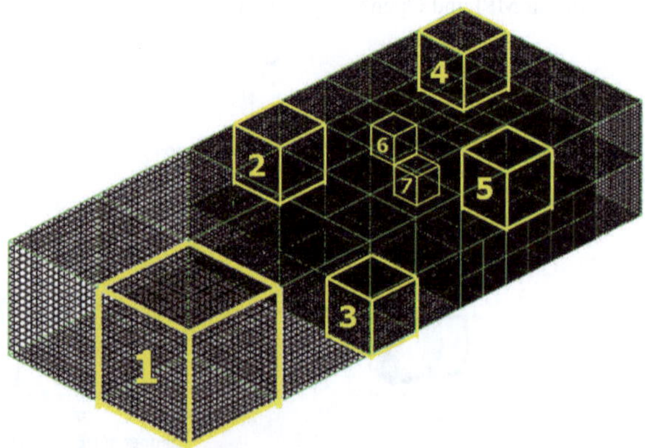

Fig. 12 Cubes assigned to the same domain (random cube address)

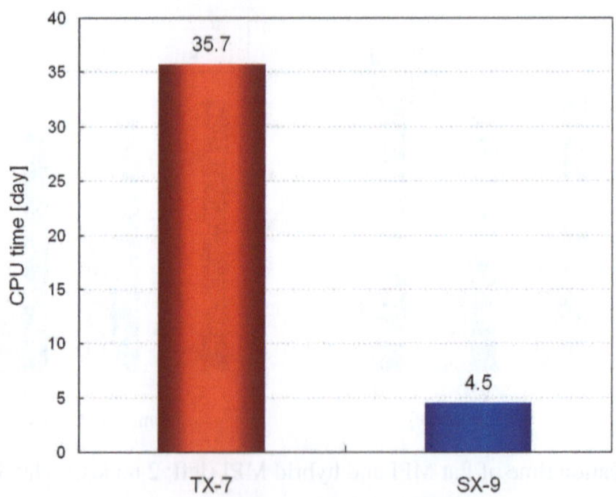

Fig. 13 Comparison of CPU time of scholar-parallel machine (TX-7) and vector-parallel machine (SX-9)

on SX-9. The computation can be fully parallelized by TX-7 with more number of cores, but there is still limitation of speed-up by pure parallelization because there are processes that cannot be parallelized.

Table 1 Computational condition for Formula 1 car model

Reynolds number	14.9×10^6
Minimum spacing	7.32×10^4
# of cells in cubes	32^3
Total # of cubes	5.930
Total # of cells	194,314,240
Convection term	3rd order upwind
Diffusion term	2nd order central
Time integration	Adams Bashforth 2nd order explicit

5 Conclusion

Large-scale incompressible flow simulation was performed on vector-parallel supercomputer of NEC SX-9 by Building-Cube Method that was based on equally-spaced Cartesian mesh method. The flow solver was tuned to maximize vector performance by implementing Red-Black coloring method and achieved 99.95% of vectorization ratio. Using Formula 1 car model with 200 million cells, the capability and parallel efficiency of large-scale computation by BCM was investigated. OpenMP showed slightly better performance than flat MPI owing to no communication latency at single node parallelization of SX-9. Both the parallel approaches showed almost the ideal scalability. On the other hand, flat MPI showed much better performance compared to hybrid MPI at multi-nodes of SX-9 due to the reduction of communication time. When 16 nodes of SX-9, which corresponds to 256 processors in total, were parallelized by flat MPI, the maximum performance achieved only 130 times improvement. The deterioration of parallel efficiency was occurred due to the random memory access for communication with neighboring cubes in each domain. This can be improved by the implementation of rearrangement technique of domain decomposition. A simple arrangement of domains according to x-coordinate of each cube has showed that nearly 160 times improvement was achieved in recent result. Finally, the large advantage of vector-parallel computers for large-scale computation of BCM was revealed. This computational efficiency of vector machines cannot be achieved by scholar-parallel computers even when more number of processors is used.

Acknowledgements Present computation was conducted by the NEC SX-9 at the Cyberscience center of Tohoku University.

References

1. TOP500 Supercomputing Sites. http://www.top500.org/, cited by June 16, 2009
2. Nakahashi, K.: High-density mesh flow computations with pre-/post-data compressions. AIAA paper, 2005–4876 (2005)
3. Nakahashi, K., Kitoh, A., Sakurai, Y.: Three-dimensional flow computations around an airfoil by building-cube method. AIAA paper, 2006–1104 (2006)
4. Takahashi, S.: Study of large scale simulation for unsteady flows. PhD dissertation, Department of Aerospace Engineering, Tohoku University (2009)
5. Ishida, T., Takahashi, S., Nakahashi, K.: Efficient and robust Cartesian mesh generation for building-cube method. J. Comput. Sci. Technol. **2**(4), 435–446 (2008)
6. Kim, J., Moin, P.: Application of a fractional-step method to incompressible Navier-Stokes equations. J. Comput. Phys. **59**, 308–323 (1985)
7. Perot, J.B.: An analysis of the fractional step method. J. Comput. Phys. **108**, 51–58 (1993)
8. Dukowicz, J.K., Dvinsky, A.: Approximate factorization as a high order splitting for the implicit incompressible flow equations. J. Comput. Phys. **102**, 336–347 (1992)
9. Kawamura, T., Kuwahara, K.: Computation of high Reynolds number flow around circular cylinder with surface roughness. AIAA paper, 84–0340 (1984)

A New Parallel SPH Method for 3D Free Surface Flows

A. Ferrari, M. Dumbser, E.F. Toro, A. Armanini

Abstract A new robust and accurate SPH scheme, developed at the University of Trento, has been applied. It is able to track correctly complex 3D free-surface flows and, even more important, also able to compute an accurate pressure field. To asses the accuracy of the new SPH scheme, a mesh-convergence study has been performed for strongly deforming free-surface in a 3D dam-break and impact-wave problem on the infrastructure provided by HLRS in Stuttgart. The parallelization of the code has been carried out using the MPI paradigm with a dynamic load-balancing strategy to improve considerably the computational efficiency of the code. Thus, simulations involving millions of particles can be run on modern massively parallel supercomputers of HLRS, obtaining a very good performance, as confirmed by a speed-up analysis. The applications in 3D consist of environmental applications. First, dam-break and impact flows against a wall have been carried out and the solutions obtained with our SPH code have been compared with either experimental data or with other numerical reference solutions. Subsequently, the focus has been on a realistic mudflow. A very complex geometry and a non-Newtonian fluid are involved. In all tests a very satisfactory agreement and results have been obtained.

1 The SPH Approach

The Smooth Particle Hydrodynamics (SPH) method is a meshless scheme based on the Lagrangian approach. The fluid is discretized by a finite set of discrete values defined at observation points. They represent material particles with physical properties, such as density, pressure, temperature, and move in space with the velocity of the fluid according to the Lagrangian approach. From the mathematical point of view, the particles are interpolation points and interact with each other through a

A. Ferrari, M. Dumbser, E.F. Toro, A. Armanini
Department of Civil and Environmental Engineering, University of Trento, Trento, Italy, e-mail: ferraria@ing.unitn.it

M. Resch et al. (eds.), *High Performance Computing on Vector Systems 2009*,
DOI 10.1007/978-3-642-03913-3_14

smoothing procedure in which a local kernel function assigns to each point a weight, based on the mutual positions of the interpolating points.

We focus on the Euler equations to simulate the free-surface non shallow water for a slightly compressible fluid, using the Tait equation to evaluate the pressure. The new SPH formulation uses an explicit third order Runge-Kutta scheme in time to improve the linear stability of the method, following a similar time evolution procedure of Oger et al. [1]. However, the explicit high order time integration is not enough to stabilize the numerical pressure field. A further stabilization is then absolutely necessary. The new key idea consists of introducing a monotone upwind flux, following directly the Vila approach [2], but only for the density equation, and for the velocity equation, the pure non-diffusive central flux without any artificial viscosity has been used. We underline that in our new approach, no artificial viscosity term must be added, which would require the careful calibration of at least two parameters depending on the test case, but only the intrinsic numerical viscosity of the monotone upwind flux in the density equation has been used to stabilize the pressure field.

2 The MPI Parallelization with Dynamic Load-Balancing

The Message Passing Interface (MPI) system is a widely used paradigm on modern massively parallel distributed-memory supercomputers. It is based on the concept that any MPI process running on its own CPU communicates via messages which other MPI processes by calling standard MPI communication subroutines. In our code, the MPI implementation uses non-blocking communication, hence decoupling each send message from the corresponding receive by the neighbour processes. The subdivision of the load-work among the CPUs is based on a spatial decomposition. In fact, within the SPH approach, the numerical domain is already organized in cells, or so-called book-keeping cells [3], which are fixed in time during the entire simulation and which contain the particles. To each CPU a non-overlapping subset of the cells is assigned. To have the best performance of the MPI communication in the code, each CPU should have approximately the same number of (fluid and solid) points to compute. Unfortunately, the Lagrangian approach of the SPH scheme introduces more complexity in the MPI implementation. In fact, an optimal distribution of the observation points to the various CPUs at the initial time does not guarantee a good efficiency during the entire simulation, since the positions of the points can change considerably in time. Each particle can move from one cell to another and subsequently it may also have to change the processor, where its data are stored. In other words, a static load-balancing approach can not assure the best reachable efficiency of the code. For this reason, the implementation of a dynamic rearrangement of the workload among the CPUs is absolutely necessary.

The dynamic load-balancing has been carried out resorting to the METIS library, a very powerful free software package for partitioning meshes and weighted graphs, proposed by Karypis and Kumar [4]. The computational domain is then partitioned so that each CPU has almost the same number of fluid particles, minimizing the MPI communication among the CPUs. The METIS package also accepts different weights for each mesh element as input data to optimize the procedure with respect to a spatially non-uniform distribution of workload, which is definitely the case for the SPH scheme, where each cell may contain a different number of particles, thus leading to a different computational effort. The weights are evaluated multiplying the number of the fluid particles per cell with the sum of all particles (fluid and solid) in the neighbour cells. Every N time-steps, METIS is called and the cells of the whole 3D computational domain are re-distributed among the CPUs to guarantee an optimal load-balancing for the new spatial configuration of the particles. This dynamic load-balancing increases considerably the efficiency of the code. The improvements in the performance depend on the numerical test case, the geometrical complexity of the computational domain and on the number of points. In particular, increasing the number of particles, the efficiency of the MPI communication improves.

A performance benchmark problem has also been computed. It consists of a dam break test case in 3D (see Sect. V) with 500,000 particles. In Fig. 1 the speed-up is plotted. It represents the computational efficiency of the code using an increasing number of CPUs (2, 4, 8, 16, 32, 64, and 128).

The computational time corresponds to the time required by the CPUs to complete the same computational job. In the graph in logarithmic scale, the optimal theoretical performance (supposing 100% MPI efficiency) is represented by the green line. The measured performance of the code produces satisfactory results, especially considering the fact that the scheme is a meshless Lagrangian method with high order Runge-Kutta time integration, which requires quite a lot of MPI communication.

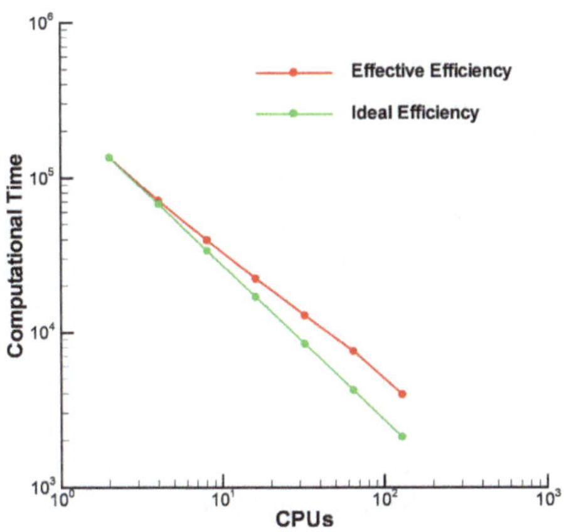

Fig. 1 Speed-up graph, comparing the measured (red) and the ideal (green) computational time from 2 to 128 CPUs

3 3D Dam Break and Impact Test Problem

This test case has been originally proposed by Colagrossi and Landrini in two space dimensions [5]. We extend it to a three-dimensional dam break problem with following impact flow against a rigid vertical wall. The initial conditions are plotted in Fig. 2 together with a 3D view of the numerical results at time $t = 2$ s. The geometrical parameters are $H = 0.6$ m, $L/H = 2$, $d/H = 5.366$ [5]. For our 3D version of this test-case, we set the channel width to $W = H = 0.6$ m.

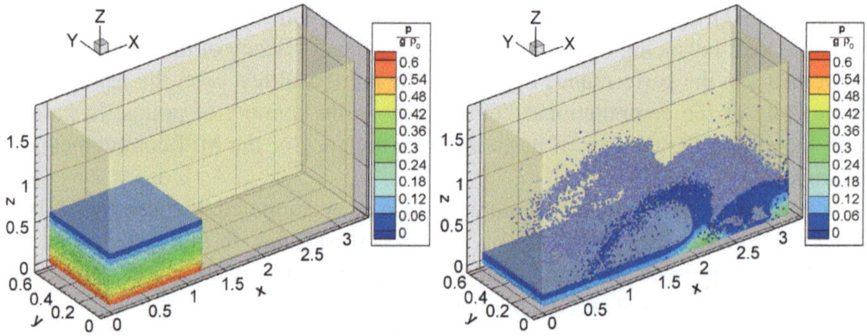

Fig. 2 The initial configuration of the 3D dam break problem (on the left) and the numerical solution at time $t = 2$ s computed by our new SPH scheme with 1,000,000 fluid particles on 128 CPUs (on the right)

The numerical solution has been computed with three different levels of refinement: 250,000, 500,000 and 1,000,000 particles to discretize the fluid. The comparison of the numerical solution has been carried out in the next section. Figure 3 shows the plots of our computational results using 1,000,000 points at various times: $t = 0.6$ s, 1.2 s, 1.5 s and 2 s. The computation has been carried out using 128 CPUs. The evolution of the pressure field for the dam break flow after the impact is plotted through the colours. As expected, our new SPH scheme does not produce any spurious oscillations in the pressure field. In particular, using the new boundary conditions, the evolution of the pressure in proximity of the solid interfaces is very smooth, without discontinuities or improper fluctuations. It is to remark that no artificial viscosity term is necessary to stabilize the new SPH scheme and no density re-initialization is required to improve the solution.

In Fig. 3 the free-surface profile before the impact (at time $t = 0.6$ s) is plotted. Moreover, the numerical velocity is compared with the classical Ritter wave front celerity [6]. In the particular case of a frictionless dam break in a long channel, the Shallow Water Equations (SWEs) can be solved analytically and the celerity of the dam break wave front is equal to UR = 2(gH)1/5, where H is the initial reservoir height in the tank. In this application, the dam break wave front celerity by Ritter results UR = 4.85 m/s. The mean SPH front velocity reaches USPH = 4.80 m/s with good agreement. Finally, in Fig. 4 a 3D view of the numerical solution computed by our SPH scheme has been plotted. It refers to the final computational instant, at time $t = 2$ s, of which the Fig. 3 represents the side-view in the x–z plane.

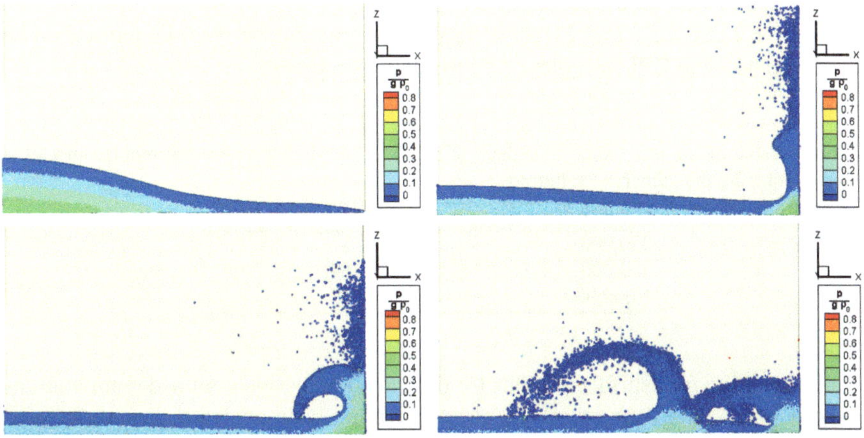

Fig. 3 The free-surface flow impacts against the vertical rigid wall. Numerical solutions computed by our new SPH scheme 1,000,000 fluid particles on 128 CPUs. The side-view is shown in the x–z plane

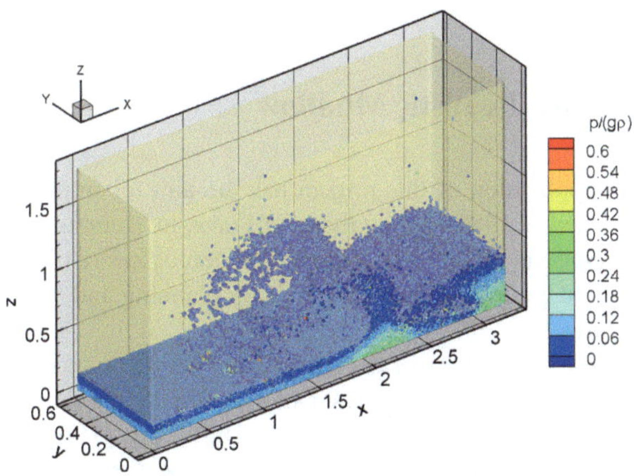

Fig. 4 A 3D view of the numerical solution at time $t = 2$ s computed by our new SPH scheme with 1,000,000 particles using 128 CPUs

4 Mesh-Convergence Test

To assess the numerical accuracy of the new 3D SPH scheme a mesh-convergence test has been carried out. Increasing the refinement of the numerical domain we have verified the effects on the solution. The analysis has been applied to three different times ($t = 1.2$ s, 1.5 s and 2.0 s). In Table 1, the number of the fluid particles with the correspondent CPUs employed by the code is reported.

Table 1 Table of the increasing refinement of the fluid and the correspondent number of CPUs, used to compute the numerical solutions

Number of particles	Number of CPUs
250,000	32
500,000	64
1,000,000	128

In Fig. 5 the free-surface profiles for the three successively refined computations are compared with each other. The drops have not been analysed, but we focused on the fluid understood as a continuum. Only on the bottom of Fig. 5 to simulate the air bubbles after the impact, some differences in the free-surface profiles are apparent. This is due to the formation of many drops and the extreme deformations of the flow. In fact, after breaking of the wave, here the free-surface is not continuous and more. Our results show that convergence has been reached after using 500,000 particles.

5 Application to a Realistic Mudflow

The challenging application of this project has the aim to simulate a catastrophic dam break, happened in 1985 on the village of Stava, in Trentino (Italy). The focus of this research consists in evaluating the hydro-geological risk for the civil society, in order to yield a correct urban and territory planning and to project adequate protection structures.

The complex geomorphology of the terrain has been reproduced very accurately (every 10 m) resorting to the digital GIS data provided by the Provincia Autonoma of Trento and the Environmental Engineering University of Trento. The tailings dams consisted of two basins on a slope and have been reconstructed following the cartography on file. The fluid derived from the processing to separate the fluorite from the rock. The released material after the collapse was high concentrated, composed of sand, slime and water. For this no relevant erosion on the bottom of the valley has been surveyed because of the flow. The mud has been discretized as a non-Newtonian fluid following a similar approach of Laigle et al. [7] and the bottom in the valley has been assumed fixed.

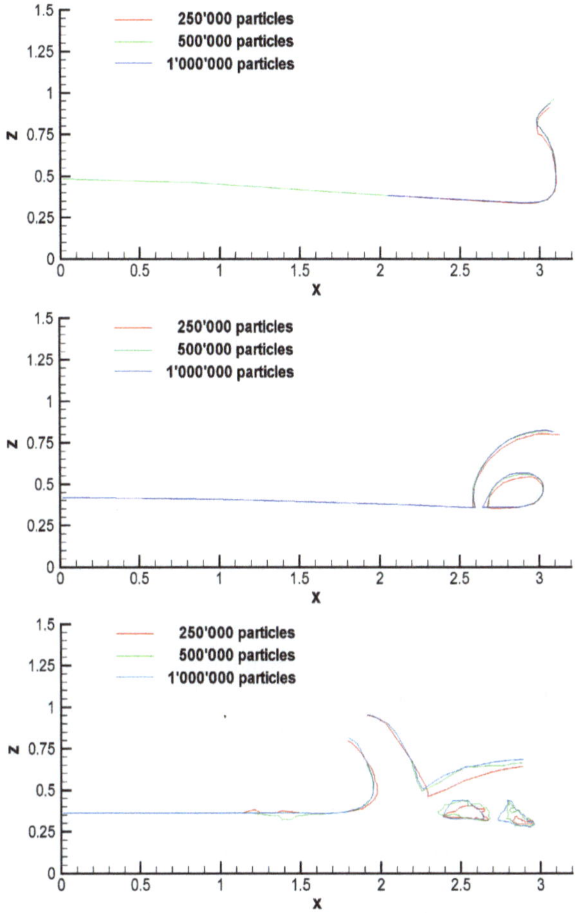

Fig. 5 Comparison of the free-surface profiles at time $t = 1.2$ s, 1.5 s and 2 s

The simulations involved millions of (fluids and solids) particles reproducing the fully three-dimensional evolution of the mudflow. Our attention has been focused on the initial instants of the phenomenon evaluating the wave propagation, where the motion follows the natural bend of the valley and the Shallow Water assumptions are not satisfied. In the following, Figs. 6 and 7 show the numerical results. They have been computed on 32 CPUs by our new SPH scheme. The solid particle to discretize the boundaries have been placed every 3 m and the initial fluid particle spacing has been imposed equal to 2 m. The numerical solutions have been compared and validated resorting to the field measurements after the dam failure and scientific contributions produced about the disaster of Stava [8].

Figure 6 consists of an aerial view of the slurry wave after the collapse together with the numerical results evaluated at $t = 30$ s and 60 s. The numerical result at time 30 s produces a larger wave front immediately after the collapse because the

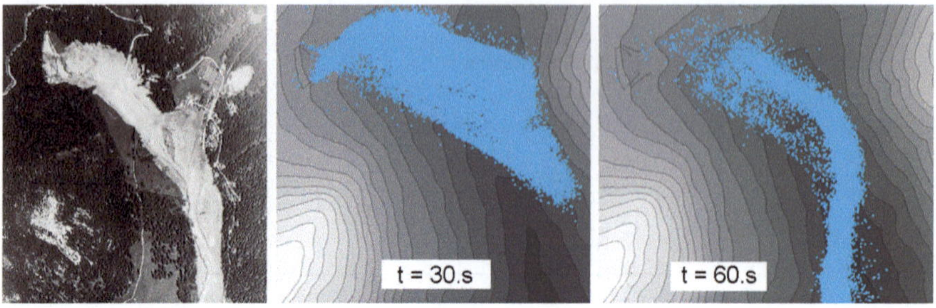

Fig. 6 Numerical solutions (on the right) evaluated at time $t = 30$ s and 60 s. They are compared with the aerial view of the valley after the collapse (on the left)

resistance to the flow due to the trees and the vegetation on the terrain (on the top of the aerial view) has not been implemented in the code. This explains the different width of the front between the aerial photography and the numerical solution. Moreover, the treatment of the collapse of the soil dams represents a problematic issue and it is not yet solved in literature, too. In this case, the mechanism of the failure and its propagation has been carried out assuming the back banks of the upper dam as rigid walls and evaluating the released muddy mass as a huge splashing wave. These assumptions are based on the documentation and field measurements on file, such i.e. the analysis carried out by Takahashi et al. [8] on the seismogram located in Cavalese. This study confirmed the testimonies asserting that the mud moved downhill at a velocity approaching 90 km/h downstream the village of Stava. After 60 s the sludge has reached the village of Stava and the mudflow is completely developed. The comparison between the aerial view and the path of the mudflow evaluated by our SPH scheme is very good. The numerical solution traces well the contour (in blue) of the overwhelmed region.

The propagation of the mudflow is represented in Fig. 7. The plots consist of the three-dimensional numerical solutions evaluated at time $t = 30$ s and 60 s. The colours denote the numerical velocity field. After the collapse of the dams at time $t = 30$ s (in Fig. 7, on the top) the slurry front propagates downstream on the complex topography of the valley impacting against the opposite mountainside. The motion is fully three-dimensional and produces adherence phenomena due to the no-slip condition of the flow at the natural bottom. The free-surface is strongly deformed with a super-elevation on the curved trajectory of the flow. In the numerical solution evaluated at time 60 s (in Fig. 7, on the bottom) the muddy wave has already reached and destroyed the village Stava. The slurry front flows into the narrow valley following the Avisio river in the direction of the village Tesero. The channelling of the mudflow produces an increase of the velocity and the width of the front reduces. From here the wave is forced to flow into the narrow valley as if it is a channel.

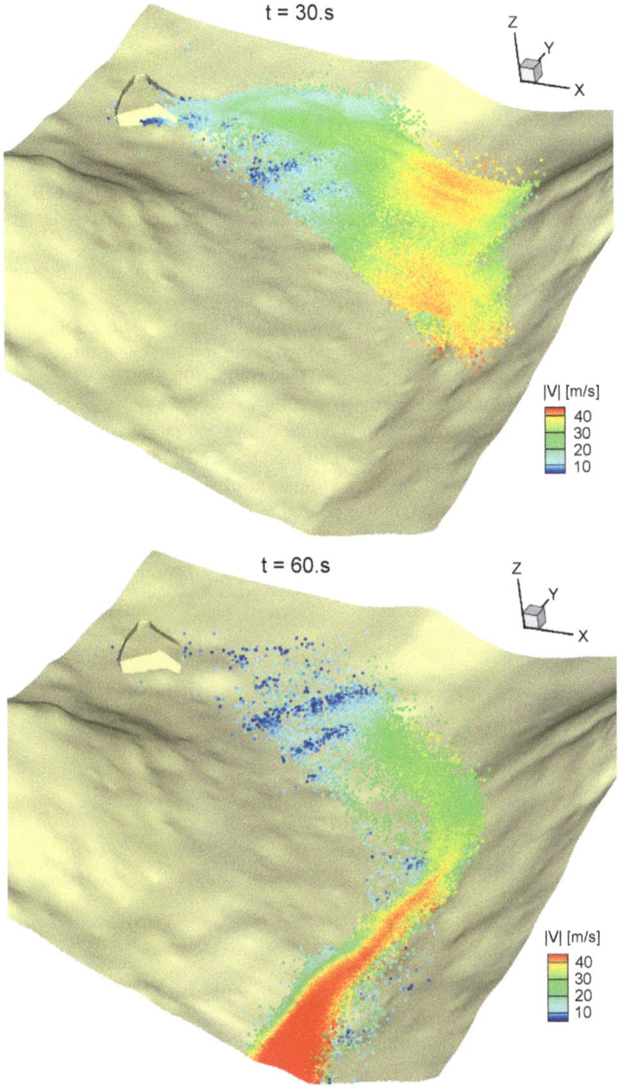

Fig. 7 Numerical solutions at time $t = 30$ s (on the top) and 60 s (on the bottom). The colour represents the velocity

Acknowledgements The work has been performed under the Project HPC-EUROPA project (application number: 1106), with the support of the European Community—Research Infrastructure Action of the FP7 "Coordination and support action" Programme.

References

1. Oger, G., et al.: Two-dimensional SPH simulations of wedge water entries. J. Comput. Phys. **213**, 803–822 (2006)
2. Vila, J.P.: On particle weighted methods and smooth particle hydrodynamics. Math. Models Methods Appl. Sci. **9**(2), 191–290 (1999)
3. Monaghan, J.J.: Simulating free surface flows with SPH. J. Comput. Phys. **110**, 399–406 (1994)
4. Karypis, G., Kumar, V.: Multilevel *k*-way partitioning scheme for irregular graphs. J. Parall. Distrib. Comput. 96–129 (1998)
5. Colagrossi, A., Landrini, M.: Numerical simulation of interfacial flows by smoothed particle hydrodynamics. J. Comput. Phys. **191**, 448–475 (2003)
6. Ritter, A.: Die Fortpflanzung der Wasserwellen. Ver. Deutsch. Ingen. Z. **36**, 947–954 (1892)
7. Laigle, D., et al.: SPH-based numerical investigation of mudflow and other complex fluid flow interactions with structures. Comput. Geosci. **11**(4), 297–306 (2007)
8. Rassegna dei contributi scientifici sul disastro della Val di Stava (Provincia di Trento), 19 luglio 1985/A review of scientific contributions on the Stava Valley disaster (Eastern Italian Alps), 19th July 1985, G. Tosatti (ed.), special vol. of GNDCI-CNR, Pitagora Editore, Bologna, 2003, ISBN 88-371-01405-2 [comprehensive compilation of 24 papers on the causes of the Stava tailings dam failure, most of which first appeared in various scientific journals; all papers are in original language (Italian, English, Japanese, German)]

V
Climate Modeling

The Agulhas System as a Prime Example for the Use of Nesting Capabilities in Ocean Modelling

Markus Scheinert, Arne Biastoch, Claus W. Böning

Abstract Today's ocean modelling faces the challenging question of how small-scale features influence ocean's larger-scale behaviour. While the latter can be simulated quite reasonably in terms of mean state and long-term variability, it lacks however some spatial precision and high frequent variability, due to a lack of horizontal resolution required to capture energetic features associated with 'meso-scale' ocean dynamics. While steady improvements of high-performance computing systems have begun to permit the simulation of such small-scale and high-frequency processes even in global configurations, enormous resources are still required for simulations over extended time spans and necessitate the development of efficient ways of implementation. Using the example of the Agulhas current system, we have used the two-way nesting technique where an eddy-resolving regional model is implanted into a global coarse resolution model to investigate the impact of small scale features in this region on the global ocean circulation variability.

1 Motivation

The large-scale variability in the oceanic circulation often involves processes occurring in the mesoscale, acting on space scales of 10–100 km. Even though in certain regions of the world ocean these motions are suspected to have a global impact, global climate studies such as the simulations for the last IPCC Assessment Report (AR5, [1]) are mainly based on coarse resolution models ($\approx 1°$[1]). Due to the complexity of the coupled ocean-atmosphere climate system and the required integration

Markus Scheinert, Arne Biastoch, Claus W. Böning
Leibniz Institut für Meereswissenschaften, Düsternbrooker Weg 20, 24106 Kiel, Germany, e-mail: mscheinert@ifm-geomar.de, abiastoch@ifm-geomar.de, cboening@ifm-geomar.de

[1] In fact, the horizontal resolution of IPCC-AR5 ocean models is between $0.3° \times 1.0°$ and $3.0° \times 4.0°$ (except for one with $0.2° \times 0.3°$ and one with $4° \times 5°$; see Table 8.1 in [2]). The bulk of the models have an average horizontal grid space of approx. 1°.

M. Resch et al. (eds.), *High Performance Computing on Vector Systems 2009*, 191
DOI 10.1007/978-3-642-03913-3_15 © Springer-Verlag Berlin Heidelberg 2010

length (o(250 years)) such models can still not be run at higher resolution. The role of mesoscale ocean processes is investigated with ocean-only models, where prescribed atmospheric boundary conditions (e.g., based on reanalysis data products) are used to simulate the ocean response over time scales of 50 to 100 years. State-of-the-art resolutions for global ocean modelling is on the order of $1/2°$ horizontal grid width ($\sim$85 km horizontal grid space at 40° S), allowing the explicit simulation of the upper part of the meso-scale spectrum (see illustration in Fig. 1). Nevertheless, some areas of the global ocean require the simulation of even finer scales of motions. Such an example is the southern tip of South Africa (see Fig. 2), a key area of the global Conveyor Belt [3] where the Indian Ocean is connected to the Atlantic Ocean via the Agulhas System. This inter-ocean transport has the potential to influence the variability of the Atlantic water masses and the global circulation of heat and salt.

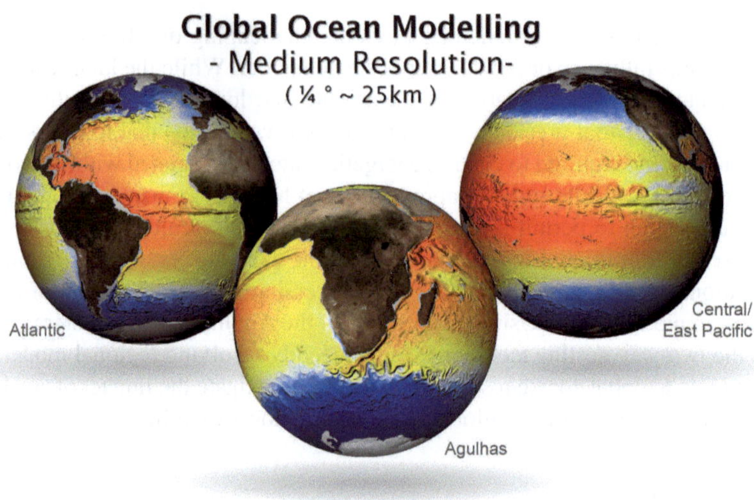

Fig. 1 Snap-shots of near surface temperature (colour indicates warm (red) and cold (blue) regions) and current speed (shaded elevation) in a global ocean simulation with a medium horizontal resolution of $1/4°$ using NEMO. The three perspective views show the Atlantic Ocean (left), the central and east Pacific (right) and the Agulhas region, where the Indian and Atlantic oceans meet. Sea ice formation is simulated as well (noticeable as white coverage at the poles)

To attain the goal of local grid refinement, several modelling techniques have been developed. Regional models with prescribed boundary conditions require a consistent prescription of the outside world, often completely dominating the interior solution of the regional model. To preserve the global context and, most importantly, to investigate the impact of local processes onto the outside world, nesting techniques have to be used: While routinely used in atmospheric models the application in the ocean modelling community was hampered for a long time due to the spurious drifts in the heat and salt budgets that limited the usefulness to a few model weeks [4]. Quite recently, a much improved mechanism for a conservative interpolation and averaging between the coarse and high-resolved grid has been developed [5]. This technique has been found to be extremely useful in ocean models by allowing an effective and consistent communication between the regional high resolution model and the global coarse resolution model acting in both directions.

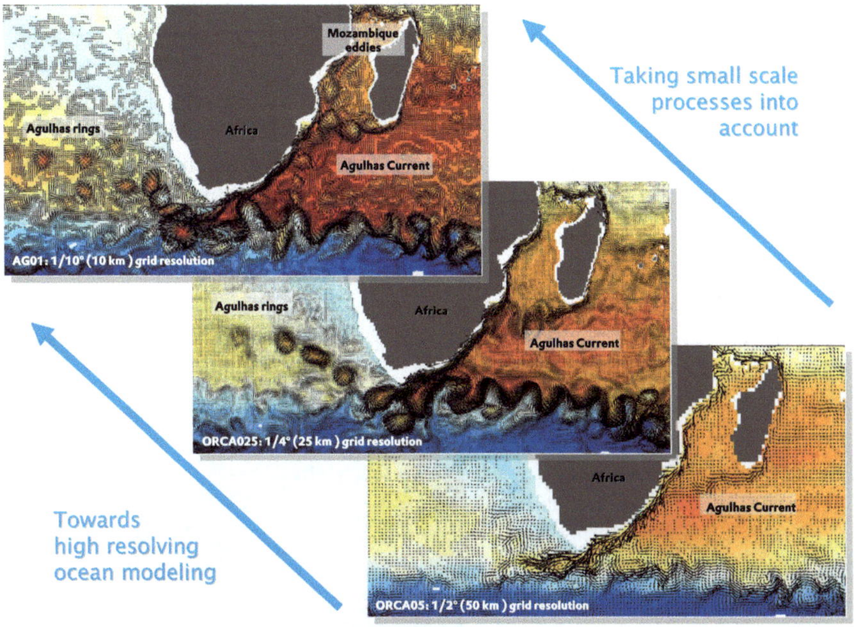

Fig. 2 Simulated near-surface temperature (colour indicates warm (red) and cold (blue) regions) and velocity (vector arrow length indicates the magnitude) in three different model configurations: ORCA05 (1/2°), ORCA025 (1/4°) and AG01 (1/10°). With increasing refinement mesoscale eddies are resolved, influencing the shape and transport variability of the large scale circulation

2 Modelling Environment

Numerical simulations performed during this project are based on global ocean-ice model configurations developed as part of the European DRAKKAR collaboration. The code basis is given by the "Nucleus for European Modelling of the Ocean" (NEMO; [6]; see http://www.nemo-ocean.eu). The ocean component (OPA) makes use of the so-called primitive equations for geophysical fluid dynamics (GFD) involving a wide scope of numerical methods: Partial differential equations for the time evolution of three-dimensional velocity, temperature and salinity fields, all discretized as finite differences; a Poisson Equation for the free ocean surface, solved by a Pre-Conjugate Gradient (PCG) method; and the equation of state, applied as a polynomial fit of the original nonlinear equation. In addition, a sea-ice model based on the viscous-plastic dynamics (LIM) is coupled to the ocean, invoking additional elliptic equations.

The code is written in FORTRAN90 and has a modular structure which allows to select specific model features during pre-compiling process. However, most parameters can be selected and set during execution using a namelist.

As the vector operation ratio easily reaches 99%, with an average vector length of more than 220, this GFD code represents an excellent candidate for the NEC SX vector machines. The parallelization is accomplished by horizontal domain decomposition while CPU/Node communication is based on MPI-1 or MPI-2. After a series of optimizations by NEC staff (cf. [7]) the performance reaches 3 (7) GFLOP/s per processor (on 16 (8) processors total) on the SX-8 (SX-9) at the HLRS computing centre.

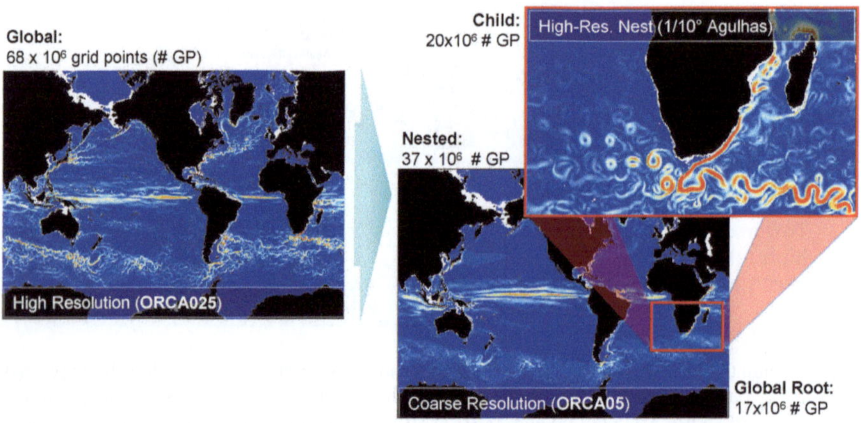

Fig. 3 Two configurations of the same ocean model based on the NEMO modelling framework: A global state-of-the-art configuration with $1/2°$ horizontal resolution (left) and a global coarser resolution model coupled with a high-resolving nest using AGRIF (right). While ORCA025 has 68×10^6 grid points (# GP) in all, the nested configuration (ORCA05 Root + Agulhas Nest) only has a combined grid point number of 37×10^6

Based on a hierarchy of coordinated experiments within the European DRAKKAR network (The DRAKKAR Group, 2007, http://www.ifremer.fr/lpo/drakkar/) model configurations were performed differing horizontal resolution and (prescribed) atmospheric conditions. The nesting technique was enabled by the implementation AGRIF (Adaptive Grid Refinement in Fortran) [8] into the NEMO code. During an automatic preprocessing step AGRIF inserts pointers into the code directing the parallel handling of both base mode and nest. The global base model (see Fig. 3) has a horizontal resolution of $1/2°$ (e.g. 50 km horizontal grid box width at 40°S) on a tripolar grid (called ORCA05) with two poles on land in the northern hemisphere to avoid a singularity in the ocean at the geographic North Pole. The vertical axis is divided into 46 unequal levels resulting in $722 \times 511 \times 46 = 17 \times 10^6$ grid points for each three dimensional variable.

The regional nested model with $1/10°$, although of quite similar size as the base model, requires a 4 times finer step to fulfill the required numerical stability criterion. This leads to a less than 20% overhead by the base model compared to the high-resolution nest alone (plus overhead by the AGRIF interpolation/averaging). An equivalent global high resolution configuration, e.g. ORCA12 ($1/12°$) would require much more computing operations (608×10^6 grid boxes and a much shorter time step) than the nested configuration.

Table 1 Comparison of configurations (names in first column) with different horizontal resolutions (second col.) using the same model. The number of grid points (#GP) are given separately for all 3 axes ($i \times j \times k$) and in the sum

		i x j x k	# GP
ORCA2	2°	182 x 149 x 31	1×10^6
ORCA05	1/2°	722 x 511 x 46	17×10^6
ORCA025	1/4°	1442 x 1021 x 46	68×10^6
ORCA12	1/12°	4322 x 3059 x 46	608×10^6

As already mentioned the coupling of base and nest model works in a two-way communication (see Fig. 4): Both models are integrated iterative, the child model gaining information from the base model along its boundaries (divided into several sub-time-steps by linear interpolation), updating the base grid along the boundaries or (every 3 base model time steps) along all base model grid points.

The atmospheric conditions are computed from forcing data sets (CORE-v1, [9]) using bulk formulae during runtime. A typical integration time of 47 years (1958–2004) requires 10,000 CPU hours for the base model alone and 35,000 CPU hours for both base and nested model together.

Such complex simulations are also expensive in terms of storage requisitions, because the complete simulated time range is needed for the variability analysis of time dependent processes. A typical experiment with the nested configuration thus easily ends up with disk consumption o(TB).

3 Scientific Achievements

Due to the high resolution and its state-of-the-art configuration the Agulhas nest has been found to simulating all mesoscale features of the current system around South Africa with great verisimilitude (see Fig. 2): A strong western boundary current, the

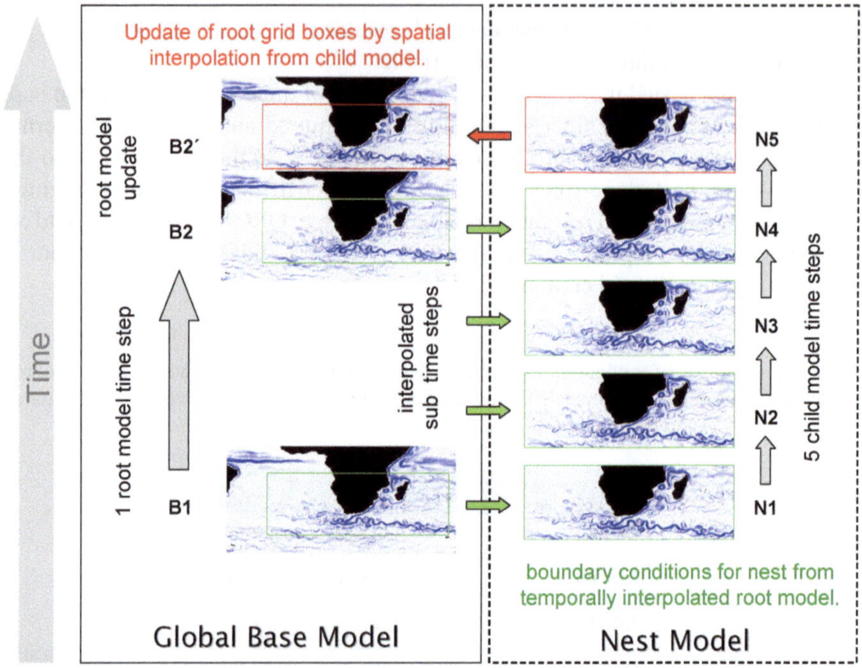

Fig. 4 Time-step and communication strategy between root and child model: The global base model is integrated forward by one step (B1 to B2). Simultaneously, the nest model is integrated with a shorter time step (e.g. N1 to N2) retrieving the boundary conditions from temporal interpolated sub-steps of the base model. After a couple of child time-steps, the base model time step is updated (B2 to B2′). The time refinement factor between base and nest in this example is 1:4

Agulhas Current [10], transports the warm and salty water southward in the Indian Ocean. South of Africa it abruptly turns back into the Indian Ocean, while shedding enormous rings, which transport heat and salt as pulsating elements into the Atlantic Ocean. The simulation also includes small-scale upstream perturbations such as Mozambique eddies as seen in satellite observations. The explicit simulation of those perturbations allows assessing their possible triggering effect on the Agulhas leakage.

The role of mesoscale processes in the net volume transfer between the Indian and Atlantic Ocean was addressed by comparing the reference simulation with an experiment in which the same global model was integrated without the high-resolution nest in the Agulhas regime [11]. The solution of this non-eddying model (ORCA05, Fig. 2) portrays the inter-oceanic exchange as a continuation of parts of the Agulhas Current as a smooth current, a solution typical of current climate models. In comparison to observations it strongly overestimates the net exchange from the Indian to the Atlantic Ocean. Only if mesoscale eddies are explicitly simulated, such as in the high-resolution nest (AG01, Fig. 2), realistic interoceanic transports appear in the solution.

What is the effect of the Agulhas Current system on the large-scale circulation in the Atlantic Ocean? By comparing the circulation in solutions with and without the high-resolution Agulhas nest it was possible to identify an intriguing contribution of the mesoscale Agulhas dynamics on decadal current fluctuations reaching far into the North Atlantic [12]. The dynamical signal originating south of Africa rapidly travels northward by boundary waves. In the tropical and sub-tropical North Atlantic the Agulhas-induced variability has similar amplitudes as the variability introduced by sub-polar deepwater formations events, a mechanism that has been known for its climatic impact and that has been extensively studied in the past. This finding highlights the importance for studying the Agulhas regime and its associated interoceanic transport as a prominent key region of the global thermohaline circulation.

4 Conclusion

The nesting technique which implants a regional ocean model with high horizontal resolution into a coarser global model has been found extremely useful for attaining a balance between an efficient usage of computing resources and scientific requirements and technical feasibility.

The fact that small-scale processes in a distinct area, e.g. mesoscale eddies in the Agulhas region, can deliberately be excluded or included just by changing the resolution in the nesting region allows to investigate not only the dynamics of these regional processes within the nesting domain, but also their impact onto the ocean circulation outside of the nest. The alternative, a high-resolution global model, would not only require much more computing time, thereby limiting the number of available sensitivity experiments. It would also simulate small scale features elsewhere in the global ocean, effectively preventing the identification of the remote effects arising from the Agulhas region.

References

1. Meehl, G.A., Covey, C., Delworth, T., Latif, M., McAvaney, B., Mitchell, J.F.B., Stouffer, R.J., Taylor, K.E.: The WCRP CMIP3 multimodel dataset—a new era in climate change research. Bull. Am. Meteorol. Soc. **88**, 1383 (2007)
2. Randall, D.A., Wood, R.A., Bony, S., Colman, R., Fichefet, T., Fyfe, J., Kattsov, V., Pitman, A., Shukla, J., Srinivasan, J., et al.: Climate models and their evaluation. Clim. Change 589–662 (2007)
3. Lutjeharms, J.R.E., de Ruijter, W.P.M.: The influence of the Agulhas Current on the adjacent coastal ocean: possible impacts of climate change. J. Mar. Syst. **7**, 321–336 (1996)
4. Fox, A.D., Maskell, S.J.: A nested primitive equation model of the Iceland-Faeroe front. J. Geophys. Res. (Oceans) **101**, 18259–18278 (1996)
5. Blayo, E., Debreu, L.: Adaptive mesh refinement for finite-difference ocean models: first experiments. J. Phys. Oceanogr. **29**, 1239–1250 (1999)

6. Madec, G.: Nemo ocean engine. Note du Pole de modélisation 27, Institut Pierre-Simon Laplace (IPSL), France, ISSN No. 1288-1619 (2008)
7. Unger, F.A., Biastoch, A.: Improvements for high resolution ocean research with NEMO. Interdiscip. Inf. Sci. **15**, 85–90 (2009)
8. Debreu, L., Vouland, C., Blayo, E.: AGRIF: Adaptive grid refinement in Fortran. Comput. Geosci. **34**, 8–13 (2008)
9. Large, W.G., Yeager, S.G.: Diurnal to decadal global forcing for ocean and sea-ice models: the data sets and flux climatologies. Technical Note TN-460+STR, NCAR, P.O. Box 3000 Boulder, Colorado 80307 (2004)
10. Lutjeharms, J.R.E.: The Agulhas Current. Springer, Berlin (2006)
11. Biastoch, A., Lutjeharms, J.R.E., Böning, C.W., Scheinert, M.: Mesoscale perturbations control inter-ocean exchange south of Africa. Geophys. Res. Lett. **35** (2008)
12. Biastoch, A., Böning, C.B., Lutjeharms, J.R.E.: Agulhas leakage dynamics affects decadal variability in Atlantic overturning circulation. Nature **456** (2008)

Seamless Simulations in Climate Variability and HPC

Keiko Takahashi, Ryo Onishi, Takeshi Sugimura, Yuya Baba, Koji Goto, Hiromitsu Fuchigami

Abstract Multi-Scale Simulator for the Geoenvironment (MSSG), which is a coupled non-hydrostatic atmosphere-ocean-land model, has been developed in the Earth simulator Center. Outline of MSSG is introduced and characteristics are presented. After optimizing computational performance of MSSG, performance analysis has been performed on the Earth Simulator. As the results of optimization, ultra high performance with MSSG has been achieved. Computational performance of the coupled non-hydrostatic atmosphere-ocean-land model has attained 52–55% of theoretical peak performance. Furthermore, results from preliminary validations including forecasting experiments are presented.

1 Introduction

Intense research effort is focused on understanding the climate/weather system using coupled atmosphere-ocean models. It is widely accepted that the most powerful tools available for assessing future weather/climate are fully coupled general circulation models. Not only interactions between atmosphere or ocean components, but also various components have been coupled with various interactive ways and influence on earth system. Getting further information on perspectives of future weather/climate and the earth system, whole of the earth system should be simulated using coupled models as much as we can.

Keiko Takahashi, Ryo Onishi, Takeshi Sugimura, Yuya Baba
The Earth Simulator Center, JAMSTC, 3173-25 Showa-machi, Kanazawa-ku, Yokohama, 236-0001, Japan

Koji Goto
NEC Corporation, 1-10 Nisshin-cho, Fuchu-shi, Tokyo, 183-5801, Japan

Hiromitsu Fuchigami
NEC Informatec Systems LTD, 3-2-1 Sakato, Takatsu-ku, Kawasaki-shi, Kanagawa, 213-0012, Japan

M. Resch et al. (eds.), *High Performance Computing on Vector Systems 2009*,
DOI 10.1007/978-3-642-03913-3_16 © Springer-Verlag Berlin Heidelberg 2010

The Earth Simulator Center have been developed Multi-Scale Simulator for the Geoenvironment (MSSG), which is a coupled non-hydrostatic atmosphere-ocean-land general circulation model, to be run on the Earth Simulator with ultra high resolution and really high performance computing architectures. When development of MSSG has been completed and various simulations are capable on the Earth Simulator, the ambition task of simulating and understanding the earth system should bring us further detail information on the Earth System. Those high qualified information for forecasting, more significant impacts might be brought by the results of simulations. Simulations with target scale shown in Fig. 1 will be planed in near future.

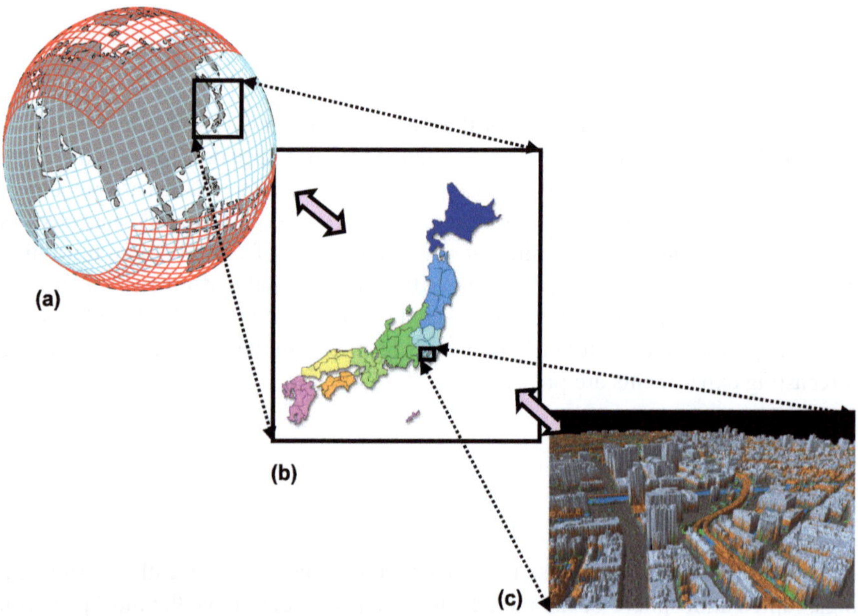

Fig. 1 (a) Yin-Yang grid system used in MSSG for the global. Each colored panel is corresponding to Yin and Yang grid, respectively. (b) Japan region is nested with two way interaction to the global. (c) As our near future target simulations, urban scale weather/climate simulations will be allowed with two-way interactions to the global/regional scale simulations. Urban topography was lent by Geographical Survey Institute

In this paper, outline of MSSG is described in Sect. 2. In Sect. 3, implementation architectures of the coupled model on the Earth Simulator are introduced. Performance analysis is performed and results are presented in Sect. 4. Preliminary validation results from simulation with the coupled non-hydrostatic atmosphere-ocean-land GCM are shown in Sect. 5.

2 Model Description

2.1 The Atmosphere Component: MSSG-A

MSSG-A is an atmosphere component of MSSG. It is compromised of the non-hydrostatic, fully compressive flux form of dynamic [1] and Smagorinsky-Lilly type parameterizations [2, 3] for subgrid scale mixing, surface fluxes [4, 5], cloud microphysics with mixed phases [6], cumulus convective processes [7, 8] and simple radiation scheme. The set of the prognostic equations is presented as follows:

$$\frac{\partial \rho'}{\partial t} + \frac{1}{G^{\frac{1}{2}} a \cos\varphi} \frac{\partial (G^{\frac{1}{2}} G^{13} \rho u)}{\partial \lambda} + \frac{1}{G^{\frac{1}{2}} a \cos\varphi} \frac{\partial (G^{\frac{1}{2}} G^{23} \cos\varphi \rho v)}{\partial \varphi} + \frac{1}{G^{\frac{1}{2}}} \frac{\partial (\rho w^*)}{\partial z^*} = 0, \tag{1}$$

$$\frac{\partial \rho u}{\partial t} + \frac{1}{G^{\frac{1}{2}} a \cos\varphi} \frac{\partial (G^{\frac{1}{2}} G^{13} P')}{\partial \lambda} = -\nabla \bullet (\rho u \mathbf{v}) + 2 f_r \rho v - 2 f_\varphi \rho w + \frac{\rho v u \tan\varphi}{a} - \frac{\rho w u}{a} + F_\lambda, \tag{2}$$

$$\frac{\partial \rho v}{\partial t} + \frac{1}{G^{\frac{1}{2}} a} \frac{\partial (G^{\frac{1}{2}} G^{23} P')}{\partial \varphi} = -\nabla \bullet (\rho v \mathbf{v}) + 2 f_\lambda \rho w - 2 f_r \rho u - \frac{\rho u u \tan\varphi}{a} - \frac{\rho w v}{a} + F_\varphi, \tag{3}$$

$$\frac{\partial \rho w}{\partial t} + \frac{1}{G^{\frac{1}{2}}} \frac{\partial P'}{\partial z^*} + \rho' \mathbf{g} = -\nabla \bullet (\rho w \mathbf{v}) + 2 f_\varphi \rho u - 2 f_\lambda \rho v + \frac{\rho u u}{a} + \frac{\rho v v}{a} + F_r, \tag{4}$$

$$\frac{\partial P'}{\partial t} + \nabla \bullet (P \mathbf{v}) + (\gamma - 1) P \nabla \bullet \mathbf{v} = (\gamma - 1) \nabla \bullet (\kappa \nabla T) + (\gamma - 1) \phi, \tag{5}$$

$$P = \rho R T, \tag{6}$$

$$\rho w^* = \frac{1}{G^{\frac{1}{2}}} (G^{\frac{1}{2}} G^{13} \rho u + G^{\frac{1}{2}} G^{23} \rho v + \rho w). \tag{7}$$

In (1)–(7), prognostic valuables are momentum $\rho \mathbf{v} = (\rho v, \rho v, \rho \omega)$, ρ' which is calculated as $\rho' = \rho - \bar{\rho}$ and P' defined by $P' = P - \bar{P}$. ρ is the density; P is the pressure; $\bar{P}$ is a constant reference pressure. $\mathbf{f}$, μ, κ, and γ are the Coriolis force, the viscosity coefficient, the diffusion coefficient, and the ratio of specific heat, respectively. F is the heat source term and the viscosity term, respectively. G is the metric term for vertical coordinate; λ is latitude; φ is longitude.

The treatment of cloud and precipitation is controlled by selecting a parameterization scheme due to horizontal resolution. For grid spacing greater than 10 km, Kain and Fritsh scheme [7, 8] is used and cloud micro physics based on mixed phase micro cloud physics [6] is used for below 5 km spacing.

Over land, the ground temperature and ground moisture are computed by using a bucket model as a simplified land model. As upper boundary condition, Rayleigh friction layer is set. For the lateral boundary condition of regional version, sponge type boundary condition [9] is used.

Regional version of the atmosphere component is utilized with one way nesting scheme by choosing the target region on the sphere, although two-way nesting is available as an option. Any large regions can be selected from the global, because both Coriolis and metric terms are introduced in the regional formulation. As another option, multiple regions are allows to be selected at the same time and computed with parallel.

2.2 The Ocean Component: MSSG-O

In the ocean component MSSG-O, the in-compressive and hydrostatic equations with the Boussinesq approximation are used based on describing in [10, 11]. The set of equations in the ocean component becomes as follows,

$$\frac{\partial c}{\partial t} = -\mathbf{v}\,\mathrm{grad}\,c + F_c, \tag{8}$$

$$\frac{\partial T}{\partial t} = -\mathbf{v}\,\mathrm{grad}\,T + F_T, \tag{9}$$

$$0 = \nabla \bullet \mathbf{v} = \left(\frac{1}{r\cos\varphi}\frac{\partial u}{\partial\lambda} + \frac{1}{r\cos\varphi}\frac{\partial(\cos\varphi v)}{\partial\varphi} + \frac{1}{r^2}\frac{\partial(r^2 w)}{\partial r}\right), \tag{10}$$

$$\frac{\partial u}{\partial t} = -\mathbf{v}\,\mathrm{grad}\,u + 2f_r v - 2f_\varphi w + \frac{vu\tan\varphi}{r} - \frac{wu}{r} - \frac{1}{\rho_0 r\cos\varphi}\frac{\partial P'}{\partial\lambda} + F_\lambda, \tag{11}$$

$$\frac{\partial v}{\partial t} = -\mathbf{v}\,\mathrm{grad}\,v + 2f_\lambda w - 2f_r u - \frac{uu\tan\varphi}{r} - \frac{wv}{r} - \frac{1}{\rho_0 r}\frac{\partial P'}{\partial\varphi} = +F_\varphi, \tag{12}$$

$$\frac{\partial w}{\partial t} = -\mathbf{v}\,\mathrm{grad}\,w + 2f_\varphi u - 2f_\lambda v + \frac{uu}{r} + \frac{vv}{r} - \frac{1}{\rho_0}\frac{\partial P'}{\partial r} - \frac{\rho'}{\rho_0}\mathbf{g} + F_r, \tag{13}$$

$$\rho = \rho(T, c, P_0), \tag{14}$$

$$\frac{d}{dr}P_0 = -\rho_0 g(r), \tag{15}$$

where the Boussinesq approximation is adopted in (9) and all variables are defined as above for the atmospheric component. In (14), UNESCO scheme [12] is used.

Smagorinsky type scheme [2, 3] is used as the subgrid-scale mixing in identical experiments with the ocean component. The level-2 turbulence closure of Mellor Yamada [13] has been also introduced to the ocean component as one of optional schemes.

In the ocean component of a regional version of MSSG-O, sponge layers are used for lateral boundary in the open ocean. The lateral boundary condition between ocean and land is defined as $\partial T/\partial t = \partial S/\partial t = 0$ and $\mathbf{v} = 0$. Bottom condition is defined by Neumann condition without vertical velocity. The upper boundary conditions are given as momentum fluxes by wind and heat fluxes from observational data of atmosphere.

2.3 Grid Configuration of MSSG

Yin-Yang grid system [14] is used both for the atmosphere-land and ocean components. Yin-Yang grid system as shown Fig. 1(a) is characterized by overlapped two panels to cover the sphere. Basically, one component grid is defined as a part of low-latitude region covered between 45N and 45S and 270 in longitude of the usual latitude-longitude grid system and the other component of the grid system is defined in the same way but in different spherical coordinates. The region covered by a panel is able to change by rotating axes of the panels.

By using Yin-Yang grid system, we can find a solution on an issue of how to avoid singular points such as the south and north poles on a latitude/longitude grid system. In addition, the advantage to enlarge the time step is used compared to conventionally utilized latitude/longitude grid system.

2.4 Differencing Schemes

In both MSSG-A and MSSG-O, the Arakawa C grid is used. The atmospheric component utilizes the terrain following vertical coordinate with Lorenz type variables distribution [15]. The ocean component uses the z-coordinate system for the vertical direction. In discretization of time, the 2nd, 3rd and 4th Runge-Kutta schemes and leap-flog schemes with Robert-Asselin time filter are available, the 3rd Runge-Kutta scheme is adopted for the atmosphere component. In this study, leap-flog schemes with Robert-Asselin time filter is used for the ocean component.

For momentum and tracer advection computations, several discretization schemes are available [16, 21]. In this study, the 5th order upwind scheme is used for the atmosphere and central difference is utilized in the ocean component.

The vertical speed of sound in the atmosphere is dominant comparing horizontal speed, because vertical discretization is tend to be finer than horizontal discretization. From those reasons, Horizontally Explicit Vertical Implicit (HEVI) scheme [21] is adopted in the atmosphere component. The speed of sound in the ocean is three times faster than it in the atmosphere, implicit method is introduced and Poisson equation (16) is solved in the method. Poisson equation is described as

$$\nabla \bullet \text{grad} P' = B, \tag{16}$$

$$B = \rho_0 \nabla \bullet \mathbf{G_v} = \frac{\rho_0}{r \cos\varphi} \frac{\partial}{\partial \lambda} G_u + \frac{\rho_0}{r \cos\varphi} \frac{\partial}{\partial \varphi} (\cos\varphi G_v) + \frac{\rho_0}{r^2} \frac{\partial}{\partial r} (r^2 G_w), \tag{17}$$

which are solved under Neumann boundary condition of $n \bullet \text{grad} P = n \bullet \mathbf{G_v}$.

No effects of over lapped grid system such as Yin-Yang grid were considered due to validations results of various benchmark experiments [16–20].

2.5 Algebraic Multigrid Method in a Poisson Solver

Algebraic Multi-Grid (AMG) method [22] is used in order to solve the Poisson equation mentioned in Sect. 2.4. AMG is well known as an optimal solution method. We used the AMG library which has been developed by Fuji Research Institute Corporation. The AMG library is characterized in terms of following points,

- AGM in the library has been developed based on aggregation-type AMG [22].
- In the library, AMG is used as a pre-conditioner in Krylov subspace algorithms.
- Incomplete LU decomposition (ILU) is adopted as a smoother in the library, which shows good computational performance even for ill-structured matrices.
- Local ILU is used for parallelization, in addition, fast convergence speed has been kept.
- Aggregation without smoothing is adopted with recalling procedure, because remarkably fast convergence has been performed by using the aggregation.

2.6 Coupling Between MSSG-A and MSSG-O

Coupling interface between the atmosphere and ocean should be taken into account to maintain a self consistent representation in the coupled model. Generally, time step in the ocean component is set longer than it of atmosphere component. The heat fluxes, moisture and momentum fluxes are computed and averaged over larger time step in MSSS-A. Those averaged fluxes are used as the upper boundary condition of the ocean component: MSSG-O. Precipitation computed in the atmosphere component is transferred to the ocean as a source term as fresh water. Sea surface temperature is defined in the most upper layer in the ocean component and is transferred to the atmospheric component as one of heat source. The SST is fixed in the atmosphere during all time steps within any large step.

3 Implementation of MSSG on the Earth Simulator

3.1 Coding Style

The coupled ocean-atmosphere model developed in the Earth Simulator Center is composed of non-hydrostatic/hydrostatic atmosphere: MSSG-A, hydrostatic/non-hydrostatic ocean: MSSG-O and simplified land components with 85,000 lines of the code. It is written in Fortran 90 and automatic allocation schemes are used in order to save memory size. MODULE features are used to keep maintainability and readability of the code. Extensions to the Message Passing Interface (MPI-2) has been used, which is tuned up for the scalability of the Earth Simulator.

3.2 Distribution Architecture and Communications

The architecture and data structures are based on domain decomposition methods. In Yin-Yang grid system, communication cost imbalance might occur by adopting one-dimensional decomposition. The case of decomposition with 16 processes is considered in Fig. 2. Each color is corresponding to each process. The number of arrows linking between different colored areas is corresponding to a mount of communication between processes. For example, in Fig. 2(a) for one-dimensional domain decomposition, black colored process called A should communicate to different colored 8 processes. In Fig. 2(b) for two dimensional decomposition, a black colored process called A communicates two processes. In Fig. 2(a), communication data size is small, in addition, the number of communications is increased. When the same number for decomposition is defined in both (a) and (b), it is clear that less amount of communication realizes in Fig. 2(b). Two-dimensional decomposition was adopted for both atmosphere-land and ocean components due to these reasons.

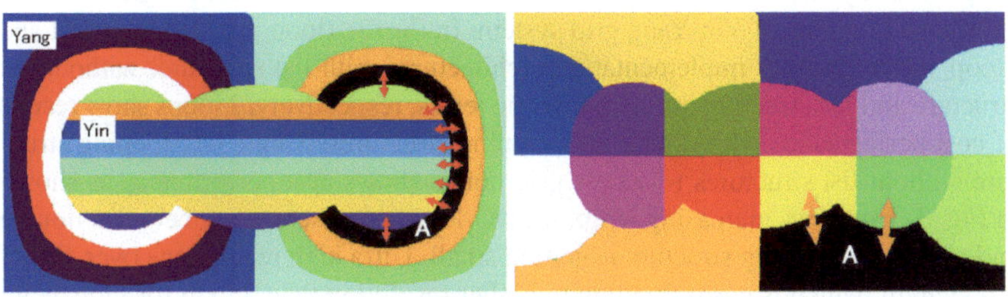

(a) One dimensional domain decomposition (b) Two dimensional domain decomposition

Fig. 2 Schematic features of domain decomposition on Yin-Yang grid system

3.3 Inter-/Intra-node Parallel Architectures and Vector Processing

Since effective parallelization and vectorization contribute to achieve high performance, the three-level parallelism which are inter-node parallel processing for distributed memory architecture, and intra-node parallel processing for shared memory architecture, and vector processing for a single processor should be utilized in order to pursuit high computational performance. Two dimensional domain decomposition was adopted to archive high performance communication by discussing in Sect. 3.2., MPI-based parallelism for inter-node is used to communicate among decomposed domains.

Microtasking of intra-node operations for shared memory architecture contributes to significant high performance, when long vector length is selected for the parallelism at a DO loop level. In order to equally share the computational load by 8 threads of microtasking, it is simple way that each microtask is mapped onto a processor and vertical layers and latitudinal grid points are parallelized with microtasking architectures. Therefore, it is necessary to high computational performance that the number of vertical layers and latitudinal grid points are required to be a multiplier of 8 to archive higher performance computation. In the cases in this paper, 32 and 40 vertical layers have been selected for the atmosphere and the ocean components, respectively.

When two-dimensional domain decomposition is used for inter-node parallelization, vector length and length of DO loops should be taken into account to be fully utilized. In this paper, two approaches in DO loops are considered in order to keep the length of DO loops. The first step to keep the length of DO loops is that both latitude and longitude direction are selected as an axis of DO loops. When the first approach is chosen, double looping structure is adopted. The second approach is that single DO looping structure is used by combining both looping axes of longitude and latitude direction.

Figure 3 shows preliminary results from computations of dynamical core with double looping or single looping structures. When single looping structure is adopted, array structures should be taken in account order to access grid point in overlapped regions of Yin-Yang grid system. In Fig. 3, 1loop_list, 1loop_nolist and 1loop_metric present implementation architectures with list structure, without list structure and list structure excepting metric terms, respectively. 2 loops show results of cost performance with double DO looping structures. Single DO looping structure without list structures to access grid points shows best performance as shown in Fig. 3. However, increasing length of DO loop, the discrepancy between double and single DO looping structure is getting small. Ultra high resolutions over 400 of loop length, which is corresponding to higher resolution than 25 km for global, are required in our simulations. Therefore, we adopted double DO looping structure, because fully high performance is expected as the same level performance of single DO looping structure and simplified coding styles are able to use in double DO looping structure.

In terms of above considering, two-dimensional decomposition, computational performance of advection terms in dynamical core is presented increasing vector

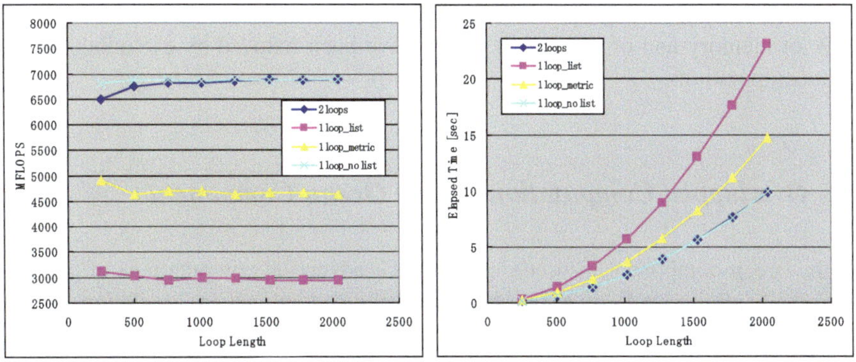

(a) (b)

Fig. 3 Cost performance for double and single DO loop structures with different horizontal resolutions. (a) and (b) show Mflops and elapsed time increasing resolutions

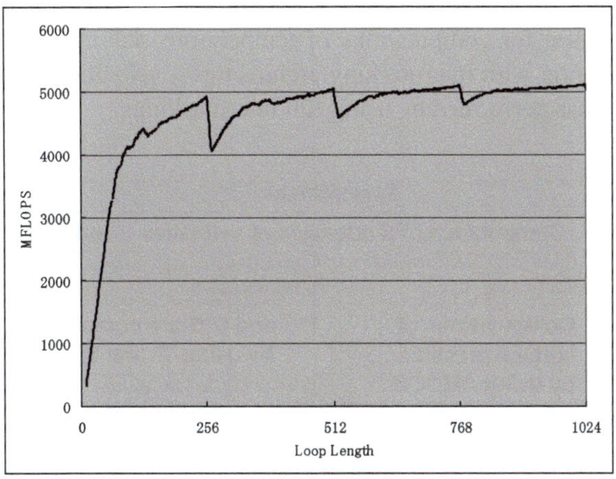

Fig. 4 Cost performance of dominant computations for advection terms in the dynamical core, increasing vector length

length in Fig. 4. As seen in Fig. 4, in a case which vector length is selected near a multiplier of 256, computational performance shows well.

3.4 Memory and Cost Reductions for Land Area in MSSG-O

By using two-dimensional domain decomposition, computations over land are not needed in the ocean component. Generally, in one-dimensional domain decomposition, masking operation is used for computations in land area. Since cost of computation for land area accounts for about 22%, those computation should be considered in order to reduce both memory and computational cost. In this paper, nodes allocated for computation in only land area are eliminated in advance of simulations. 22% of memory and of cost performance have been reduced by excluding redundant computations.

3.5 Overlapped Computations in the Ocean Component

In the ocean component, procedures of AMG, advection computations of temperature and salinity, and computation of density are dominant in simulations of the ocean component. Since procedures of AMG are able to be performed independently of advection and density computations. Microtasking of intra-node operations for shared memory architecture is used for parallelization of AMG procedures and advection computations. Figure 5 shows an outline of parallelization with microtasking architecture. After using eight microtasking for computations of velocities u and v, one microtask is used for Poisson solver with AMGCG and seven microtasks are used for computations of temperature, salinity and density. After parallel computation with microtasking architectures, velocities are computed and communication has performed by using full microtaskings.

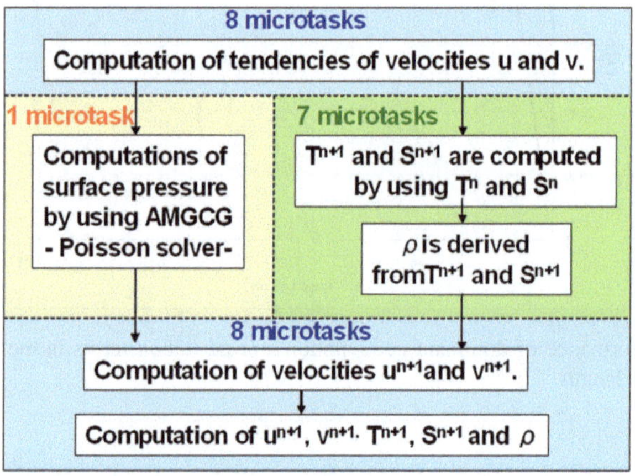

Fig. 5 Schematic figure of flow chart for parallelization with microtasking architectures in the ocean component

3.6 Coupling Scheme with High Computational Performance in MSSG

In most of coupled models which was publicized, boundary conditions are replaced by interface conditions linking between the component models. The interface between atmosphere and ocean should be taken into account to maintain a self consistent representation in a model. To remove causative artificial factors of the inconsistent through coupling, two approaches are considered for coupling schemes. For the one of them, individual component can run independently and communicate surface variables after integration during the same duration. In this framework, if each component is coupled by fully parallelized interface, computational performance efficiency has been improved. However, it is difficult to find the most suitable number of nodes used for each component without try and error experiments. Furthermore, parallel computational efficiency is restricted by dividing nodes for each component and communication cost imbalance between the atmosphere and ocean components can not be avoided.

If each component is fully optimized and can be implemented with high parallelization on whole nodes of the Earth Simulator, sequential coupling shows much higher computational efficiency. In this coupling scheme, the inconsistent through coupling should be avoided. After integration with the atmosphere component, averaged fluxes during the integration have to be stored in memory until finishing of the following integration with the ocean component. Since the atmosphere and ocean components are assumed to be fully optimized and parallelized in this study, the serial coupling scheme is adopted as shown in Fig. 6.

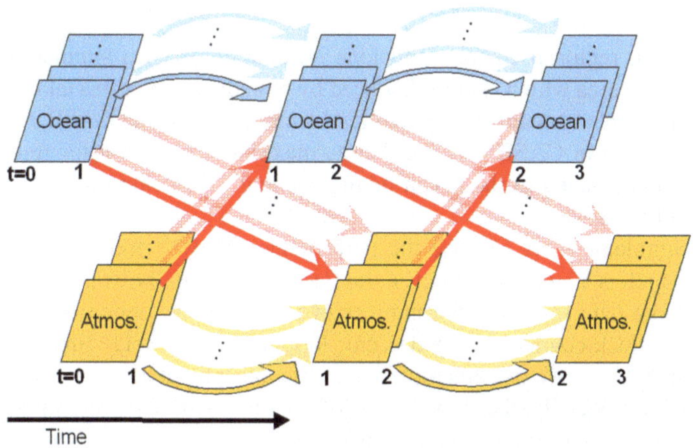

Fig. 6 Schematic figure of serial coupling scheme between the atmosphere and ocean components. Arrows show the direction of data flow

4 Computational Performance on MSSG on the Earth Simulator

4.1 Performance and Scalability

Since the developed coupled ocean-atmosphere model can be run under various conditions, several cases are selected up in order to measure computational performance as follows.

CASE 1: the developed coupled ocean-atmosphere-land model with 1.5 km resolution in horizontal direction and 72 vertical layers for Japan region.
CASE 2: the developed stand-alone atmosphere-land component for the global with 2.26 km horizontal resolution and 32 vertical layers.
CASE 3: the developed stand-alone atmosphere-land component for Japan region with 1.3 km horizontal resolution and 32 vertical layers.
CASE 4: the developed stand-alone ocean component with 1.4 km resolution in horizontal direction and 40 vertical layers for the North Pacific basin and region between the equator and 30°S.

In all cases of CASEs of 1, 2 and 3, cloud microphysics is used for handling both non-convective and convective clouds and precipitation in the atmospheric component including the coupled model. The atmospheric component of the developed coupled model performs as a cloud resolving model. In CASEs 1 and 3, horizontal resolution is decided by the reason that non-hydrostatic physical phenomena in atmospheric component such as rain band is well represented with horizontal resolution less than 5 km. In CASE 2, horizontal resolution and the number of vertical layers are decided by the restriction of total memory size of the Earth Simulator used by global simulation with the developed stand-alone atmospheric model. The Japan region used in CASE 1 and CASE 3 is defined referring to in hazard region for 72 hours forecasting of typhoons tracking used by Japan Meteorological Agency.

In CASE 4, region used for standalone oceanic component simulations is defined as enough area which is not influenced by boundary conditions taking account for regional coupled ocean-atmosphere-land simulation such as CASE 1. The horizontal resolution in CASE 4 is decided as maximum resolution for representing hydrostatic phenomena in the ocean taking account to matching to the horizontal resolution of CASE 1.

Earth Simulator users can use the performance analysis tool FTRACE (Flow TRACE) which is a built-in counter in the Earth Simulator. By using FTRACE, we can obtain the data such that the number of floating-point operations and vector instructions, clock counts, averaged vector loop length and delay time due to out-of-cache operations. We use this tool to measure the computational performance of each CASEs. Especially, the flops values of the all CASEs are determined on the basis of the performance information output for each MPI process. Each flop value can be derived as the total number of floating point operations for all the processors divided by the maximum elapsed time.

Computational performances for all CASEs are shown in Table 1.

Table 1 is shown for each CASE with various processor configurations ranging from 256 nodes by 8 intra-node processors to 512 nodes by 8 processors. The coupled non-hydrostatic atmosphere-ocean-land model has achieved a very excellent overall sustained performance of 17.07 Tflops, which is 52.1% of the theoretical peak performance on 512 nodes of the Earth Simulator. For both of stand alone non-hydrostatic atmosphere and ocean components have attained well performance on the Earth Simulator. Especially, in global simulation with stand alone non-hydrostatic atmosphere component shows that sustained performance presents

Table 1 Computational performance on the Earth Simulator

CASE	TPN	TAP	grid pts	Mflops/AP	Vector length	V.OP ratio	Tflops	Peak ratio	Parallel efficiency	Speedup
1	512	4096	3,662,807,040	4166.7	229	99.3%	17.07	52.1%	90.0%	461.0
	384	3072		4273.8	229	99.3%	13.13	53.4%	92.3%	354.6
	256	2048		4401.9	229	99.3%	9.02	55.0%	94.8%	242.6
2	512	4096	2,713,190,400	4575.2	228	99.5%	18.74	57.2%	93.6%	479.1
	384	3072		4606.1	228	99.5%	14.15	57.6%	95.1%	365.2
	256	2048		4692.4	228	99.5%	9.61	58.7%	96.7%	247.5
3	512	4096	2,713,190,400	4340.8	229	99.4%	17.78	54.3%	90.7%	464.4
	384	3072		4401.0	229	99.4%	13.52	55.0%	92.9%	356.6
	256	2048		4560.5	229	99.4%	9.34	57.0%	95.1%	243.5
4	498	3984	4,718,592,000	3629.3	240	99.3%	14.46	45.4%	80.6%	401.3
	398	3184		3568.5	240	99.3%	11.36	44.6%	83.8%	333.7
	303	2424		3986.8	240	99.3%	9.66	49.8%	87.2%	264.2
	207	1656		4234.3	240	99.3%	7.01	52.9%	90.9%	188.2

CASE identifies the name of the above cases; TPN is the total number of nodes; TAP is the total number of arithmetic processors; grid pts is the number of grid points in the each CASE; Mflops/AP is the corresponding megaflops per an arithmetic processor; Vector length is averaged length of vector processing; V. OP ratio is the vector operation ratio; Tflops is the total telaflops sustained over the duration, exclusive of I/O; Peak ratio is the percentage of total telaflops to the theoretical peak performance; Parallel efficiency and Speedup are measured by degradation in elapse time relative to single arithmetic processor, respectively

18.74 Tflops, which is 57.2% of the theoretical peak performance on 512 nodes. This result shows that the well sustained performance is obtained not only for the coupled model but also both stand alone atmosphere and ocean components on wide range of system configurations of the Earth Simulator.

4.2 Cost Balance and Communication Cost

In CASE 1: the coupled non-hydrostatic atmosphere-ocean-land GCM, cost of each component and communication between atmosphere and ocean components are shown in Table 1. Horizontal resolution of the atmospheric component is same as it of the oceanic component, because total amount of fluxes throughout communications between the atmosphere and ocean components should be conserved during whole integration. In simulations with atmosphere horizontal resolution same as it of ocean component such as CASE 1, the same domain was distributed to each node. Therefore, the shared memory architecture within each node is used for communication of fluxes between atmosphere and ocean components. Table 2 shows that low cost of communication due to the shared memory has been achieved. Furthermore, Table 2 suggests that optimization of an atmospheric component plays an important role in order to realize high performance computation with a coupled atmosphere-ocean model under the condition of the same atmosphere horizontal resolution as it of an ocean component.

Table 2 Cost balance on 512 nodes of the Earth Simulator in CASE 1

Component	Elapsed time (sec)	Ratio to total elapsed time
Atmosphere component	460.641	97.09%
Ocean component	13.548	2.86%
Data exchanging for coupling	0.256	0.05%
Total elapsed time	474.445	–

Table 3 shows cost balance of each processes in CASE 2: the atmosphere component: MSSG-A simulations for global. Cost of communications in physical processes is relatively high as shown in Table 3. It suggests there might be cost imbalance. The cost imbalance reveled in cloud microphysics processes. We examined introducing a balancer scheme, which can change decomposition region to be divided to each node, to smooth cost imbalance among processes, however, cost imbalance has not been improved. It is still an open problem to improve cost performance in physical processes.

Table 3 Cost balance on 512 nodes of the Earth Simulator in CASE 2

Processes/Schemes	Elapsed time (sec)	Ratio to total elapsed time
Whole atmospheric component	1773.36	
Dynamical core processes	1435.39	80.94%
Physics processes	337.97	19.06%
Processes/Schemes	Ratio to total elapsed time of physics processes	Ratio to total elapsed time of physics processes
Cloud microphysics	67.20%	12.81%
Surface fluxes schemes	1.80%	0.34%
Subfield variables	8.35%	1.59%
Radiation schemes	3.02%	0.58%
Land processes	1.71%	0.33%
Others	1.08%	0.21%
Communications	16.85%	3.21%

4.3 *Efficiency of Overlapped Computation in the Oceanic Component*

In CASE 4 with the ocean component: MSSG-O, overlapped computation is introduced as mentioned in Sect. 3.4. The efficiency is shown in Table 4. After introducing it, total elapsed time decreases with about 20%. The parallelization is implemented by using microtasking architecture within each node, communication cost of data transfer among each task can be neglect by using shared memory in a node.

Table 4 Cost efficiency with overlapped computation in CASE 4 on 498 nodes of the Earth Simulator

Schemes	Overlapped computation		Computation without overlap			
	Total elapsed time (sec)	Ratio to total elapsed time	Total elapsed time (sec)		Ratio to total elapsed time	
Whole ocean component	576.281	–		688.251		–
Two-dimensional Poison solver	246.641	42.80%	350.957	247.323	50.99%	35.94%
Computation of T, S and ρ				103.634		15.06%

T is temperature; S is salinity; and ρ shows density in the ocean

5 Simulation Results

5.1 Global Simulation with MSSG-A

Sophisticated coding style of MSSG such as memory management, information communication schemes, optimizations on the ES, enable us huge simulation with MSSG. Figure 4 shows precipitation distribution snapshot of MSSG-A simulation results with 1.9 km for horizontal and 32 vertical layers. One week integration was executed with the atmospheric component. Initialized data was interpolated at 00UTC08Aug2003 from Grid Point Value (GPV) data provided by Japan Meteorological Business Support Center. Sea surface data was also made by GPV data at 00UTC08Aug2003 and fixed during the simulation. Precipitation distribution for global has been presented in Fig. 7. Figure shows averaged precipitation one hour before 00UTC10Aug2003. The unit is mm per hour. Precipitation distribution has been brought by working of cloud microphysics and is comparable to observation data. In this simulation, diurnal cycle of precipitation in Indonesian region and fine structure of fronts are captured.

Regional validation has been performed with one way nesting from 5.5 km global simulation. Horizontal resolution and vertical layers were set the same condition as it of CASE 3. Initialized data was interpolated by using GPV data at 00UTC08Aug2003 provided by Japan Meteorological Business Support Center Boundary condition was made by interpolation the above simulation with 5.5 km horizontal resolution. Sea surface temperature was also fixed to data at 00UTC08Aug2003 during the simulation. 72-hours integration has been performed. Figure 8 shows the result after 72 hours integration. Meso–γ scale disturbance such as rain band might be captured in the simulations.

5.2 Stand Alone Oceanic Component

In MSSG-O, as validation simulation, 15-years integration with 11 km horizontal resolution and 40 vertical layers has been executed for the North Pacific basin and

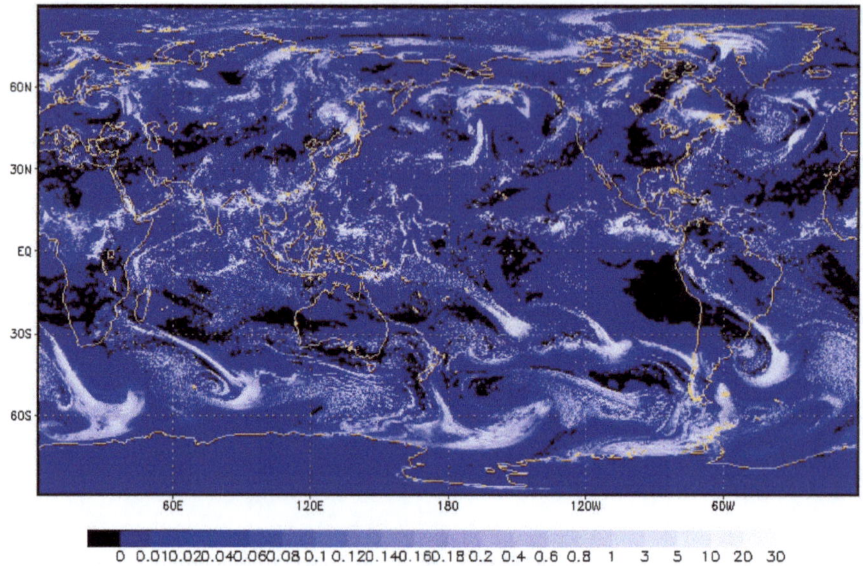

Fig. 7 Global precipitation distribution with MSSG-A

region between the equator and 30°S in CASE 4 Surface heat fluxes and boundary data are computed from climatological data provided by World Ocean Atlas (WOA). Momentum fluxes are obtained by interpolating from climatological data by NCAR. Figure 9 shows the snap shot in April after 15 years integration. Figure 9(a) shows temperature distribution at 15 m depth from the surface, which is corresponding to the second layer from the surface. Figure 9(b) shows distribution of absolute value of horizontal velocity at 105 m depth. Eddy resolved distributions have been recognized in both Fig. 9(a) and (b).

5.3 Prediction of Typhoon Tracking with MSSG

Atmosphere-Ocean Coupled Model: MSSG was tested by 120 hours forecasting experiments for typhoon tracking of ETAU during the term from 15UTC06Aug2003 to 15UTC11Aug2003. 120 hours forecasting has performed by using boundary condition with results of forecasting with global MSSG-A. In the region surrounded by 19.582°N–49.443°N, 123.082°E–152.943°E, MSSG-A and MSSG-O are coupled with 2.78 km for horizontal. For vertical layers, 32 layers and 44 layers were used for the atmosphere and ocean components, respectively.

Initial data of atmospheric was also given by GPV data with interpolation at 15UTC06Aug2003 for the atmosphere. For the ocean component, further 24 hours integration has been performed. After these global atmospheric simulations with 5.5 km horizontal resolution and 32 vertical layers, regional boundary data for

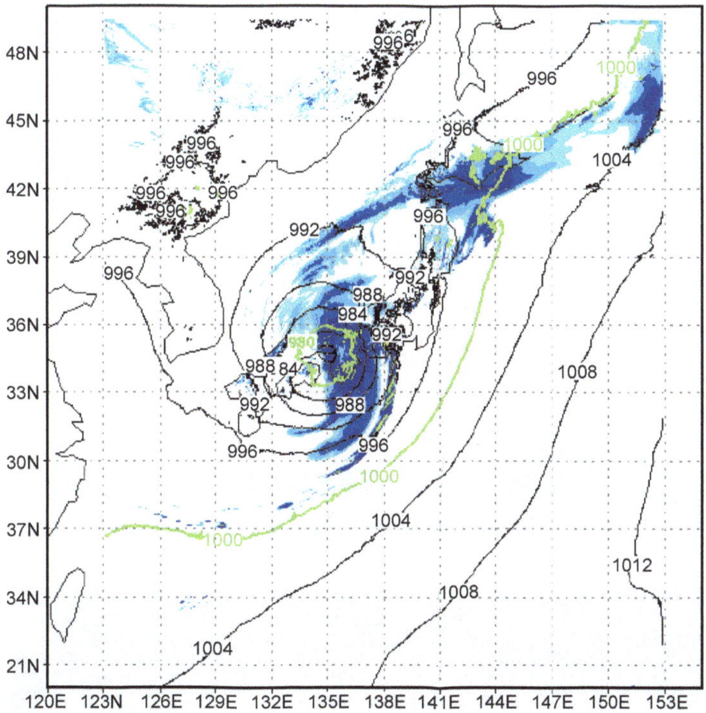

Fig. 8 Regional validation results with the atmospheric component. Colored distribution shows the precipitation (mm/hour)

Japanese region has been obtained from GPV data. In the oceanic component, initial data at forecast beginning date of 15UTC06Aug2003 was obtained by doing 10 days spin up integration from 27th July in 2003 based on July climatological data of previous 15 years integration. During the 10 days spin up integration, surface boundary data is given by 6 hourly data by NCAR. Outside of the focused Japan region, global atmosphere simulation with 5.5 km horizontal resolution performed and its results were used as lateral boundary condition of the atmosphere component in Japan region. Lateral condition of the ocean component in Japan region was given by results of climatological data from previous 15 years integration. Coupling was done without any flux correction.

Results are presented in Fig. 10. Observation data presented (a) in Fig. 10 is opened it to the public as 'best tracking' by Japan Meteorological Agency. Simulated track of ETAU is comparable with other forecasting results. Figure 11(A)–(F) shows time series results of 1 hour averaged forecasting with the coupled model in Japan region. Distribution of blue gradation shows precipitation distribution. Fine structure like a rain band has been represented in (A). Those distribution structure showed drastic change as a typhoon attacked Japan and went through Japan. In the ocean, SST response to a typhoon was simulated in (B)–(F). Oscillation due to dis-

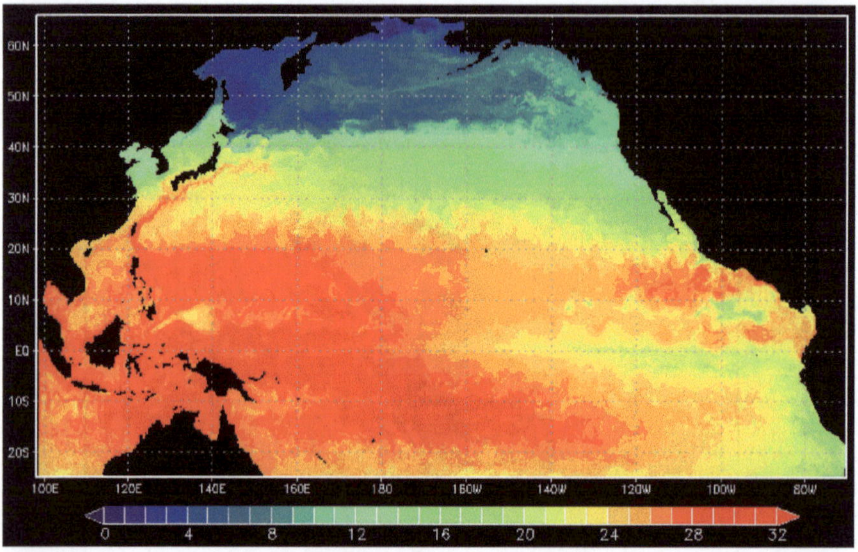

(a)

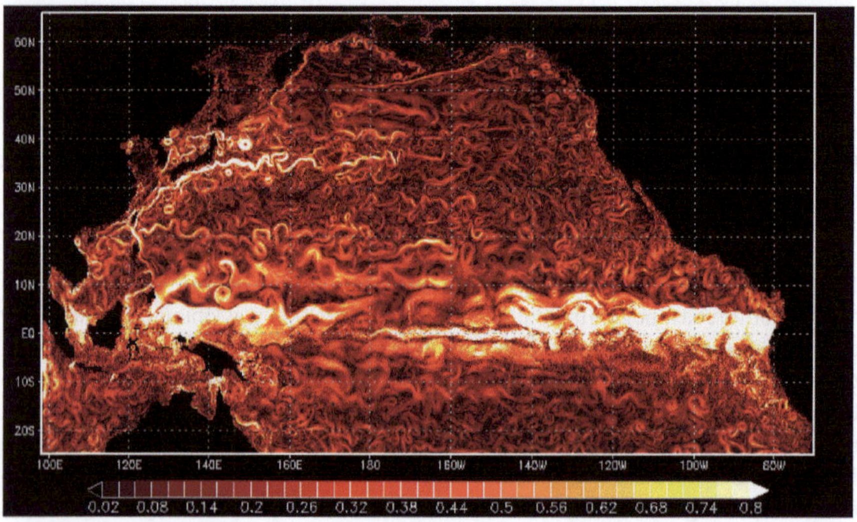

(b)

Fig. 9 Snap shot results from regional simulations with the ocean component after 15 years integration. (a) Sea surface temperature (°C) at 15 m depth from surface. (b) Absolute value distribution of horizontal velocity (m/sec) at 105 m depth. Color contour in (b) is used referring to the color bar in http://ecco.jpl.nasa.gov/cube_shere

turbance of a typhoon was recognized in not only SST but also vertical velocity and Kuroshio. Detailed analysis on eye core structure or ocean responses is still going on.

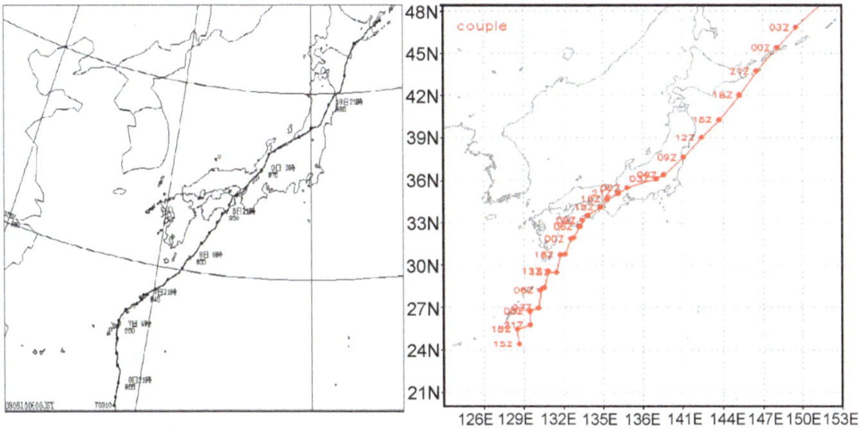

Fig. 10 Results from tracking forecast of typhoon ETAU. (a) Best tracking by JMA, (b) Simulation results from 72 hours forecasting of ETAU

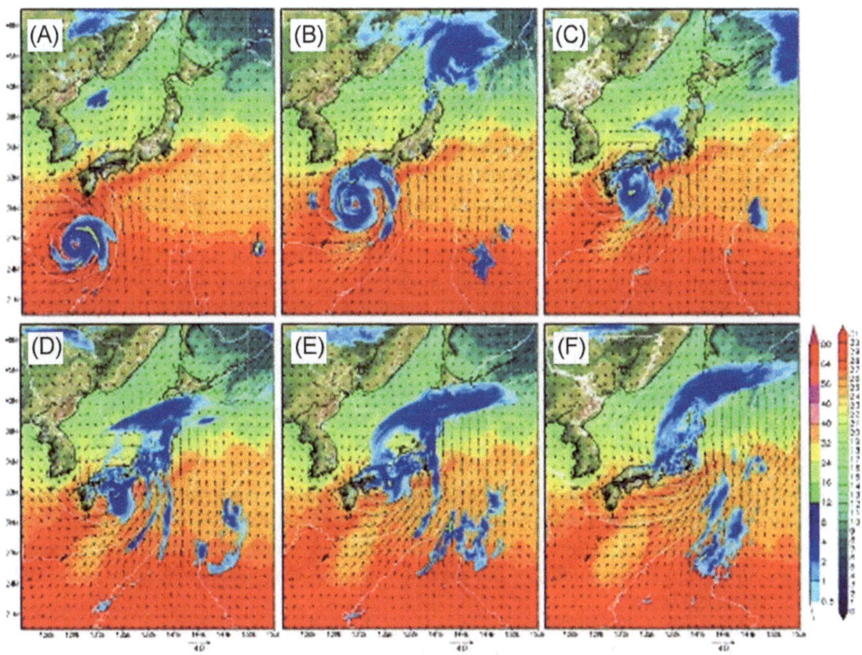

Fig. 11 Precipitation distribution (mm/h), wind velocity with black allow and SST distribution during typhoon ETAU attacked Japan region. Left-hand side color bar shows volume of precipitation and right-hand side color bar presents SST temperature

6 Conclusions and Perspectives

The developments of the coupled non-hydrostatic atmosphere-ocean-land general circulation model were successfully completed with the high sustained performance of 17.07 Tflops which is corresponding to 52.1% of peak performance with the full utilization of 512 nodes of the Earth Simulator. The components of the coupled model showed high sustained performance as well as the coupled model. These results encourage us to start various simulations with the coupled non-hydrostatic atmosphere-ocean-land general circulation model. As challenging issues, more various forecasting experiments is going to be performed and more longer integration will be executed in order to estimate the accuracy of forecasting.

References

1. Satomura, T., Akiba, S.: Development of high-precision nonhydrostatic atmospheric model (1): Governing equations. Annu. Disas. Prev. Res. Inst. Kyoto Univ. B **46**, 331–336 (2003)
2. Lilly, D.K.: On the numerical simulation of buoyant convection. Tellus **14**, 148–172 (1962)
3. Smagorinsky, J., Manabe, S., Holloway, J.L. Jr.: Numerical results from a nine level general circulation model of the atmosphere. Mon. Weather Rev. **93**, 727–768 (1965)
4. Zhang, D., Anthes, R.A.: A high-resolution model of the planetary boundary layer—sensitivity tests and comparisons with SESAME-79 data. J. Appl. Meteorol. **21**, 1594–1609 (1982)
5. Blackadar, A.K.: High resolution models of the planetary boundary layer. In: Pfafflin, Ziegler (eds.) Advances in Environmental Science and Engineering, vol. 1, pp. 50–85. Gordon and Breach, Newark (1979)
6. Reisner, J., Ramussen, R.J., Bruintjes, R.T.: Explicit forecasting of supercooled liquid water in winter storms using the MM5 mesoscale model. Q. J. R. Meteorol. Soc., (1998)
7. Kain, J.S., Fritsch, J.M.: Convective parameterization for mesoscale models: the Kain-Fritsch scheme. In: The Representation of Cumulus Convection in Numerical Models of the Atmosphere, Meteor. Monogr., vol. 46, pp. 165–170. Am. Meteorol. Soc. (1993)
8. Fritsch, J.M., Chappell, C.F.: Numerical prediction of convectively driven mesoscale pressure systems, Part I: Convective parameterization. J. Atmos. Sci. **37**, 1722–1733 (1980)
9. Davies, H.C.: A lateral boundary formulation for multi-level prediction models. Q. J. R. Meteorol. Soc. **102**, 405–418 (1976)
10. Marshall, J., Hill, C., Perelman, L., Adcroft, A.: Hydrostatic, quasi-hydrostatic, and nonhydrostatic ocean modeling. J. Geophys. Res. **102**, 5733–5752 (1997)
11. Marshall, J., Adcroft, A., Hill, C., Perelman, L., Heisey, C.: A finite-volume, incompressible Navier Stokes model for studies of the ocean on parallel computers. J. Geophys. Res. **102**, 5753–5766 (1997)
12. Gill, A.: Atmosphere-Ocean Dynamics. Academic Press, San Diego (1982)
13. Mellor, G.L., Yamada, T.: A hierarchy of turbulence closure models for planetary boundary layers. J. Atmos. Sci. **31**, 1791–1806 (1974)
14. Kageyama, A., Sato, T.: The Yin-Yang grid: an overset grid in spherical geometry. Geochem. Geophys. Geosyst. **5**, Q09005 (2004). doi:10.1029/2004GC000734
15. Gal-Chen, T., Somerville, R.C.J.: On the use of a coordinate transformation for the solution of the Navier-Stokes equations. J. Comput. Phys. **17**, 209–228 (1975)
16. Peng, X., Xiao, F., Takahashi, K., Yabe, T.: CIP transport in meteorological models. JSME Int. J., Ser. B **47**(4), 725–734 (2004)
17. Komine, K.: Validation results from non-hydrostatic atmospheric simulations. In: 2005 SIAM Conference on Computational Science and Engineering, Florida, USA, Feb 2005

18. Ohdaira, M., Takahashi, K., Watanabe, K.: Validation for the solution of shallow water equations in spherical geometry with overset grid system in spherical geometry. In: The 2004 Workshop on the Solution of Partial Differential Equations on the Sphere, pp. 71, Yokohama, July 2004
19. Takahashi, K., et al.: In: Proc. 7th International Conference on High Performance Computing and Grid in Asia Pacific Region, pp. 487 (2004)
20. Takahashi, K., et al.: Non-hydrostatic atmospheric GCM development and its computational performance. http://www.ecmwf.int/newsevents/meetings/workshops/2004/high_performance_computing-11th/presentations.html
21. Wicker, L.J., Skamarock, W.C.: Time-splitting methods for elastic models using forward time schemes. Mon. Weather Rev. **130**, 2088–2097 (2002)
22. Stuben, K.: A review of algebraic multigrid. GMD Report 96 (1999)

VI
Computational Physics

Construction of Vibration Table in an Extended World for Safety Assessment of Nuclear Power Plants

Tomonori Yamada, Fumimasa Araya

Abstract Computational issues of the seismic simulation of nuclear power plants are discussed in this chapter. The safety requirement of nuclear power plant is high enough to conduct real experiments to evaluate the structural integrity of mechanical components. However, such experiments are performed on independent sets of components because the ability of experimental facilities is limited. Hence, we are constructing a vibration table in an extended world for simulation. In our vibration table, the balancing domain decomposition method is adopted to reduce the computation cost of large mechanical components. Because the computation cost of balancing domain decomposition method depends much on the number of subdomains, a prediction methodology of optimal number of subdomains is introduced. Numerical validation is performed with a component of an actual nuclear power plant.

1 Introduction

The importance of nuclear power generation for ensuring national energy security is widely acknowledged, and it is also critical to guarantee the safety of nuclear power plants. The seismic safety of nuclear power plants has attracted great concerns after the introduction of new regulatory guidelines for the seismic design of nuclear power plants in Japan and also after recent strong earthquakes. The Nuclear Safety Commission of Japan summarized the research activities on the seismic safety of nuclear power plants and laid stress on the importance of these research activities [1]. When the Japanese Ministry of Economy, Trade and Industry announced the US-Japan Joint Nuclear Energy Action Plan, seismic analysis technology was

---

Tomonori Yamada, Fumimasa Araya
Center for Computational Science and e-Systems, Japan Atomic Energy Agency, 6-9-3 Higashiueno, Taito Tokyo, 110-0015, Japan e-mail: yamada.tomonori@jaea.go.jp, araya.fumimasa@jaea.go.jp

M. Resch et al. (eds.), *High Performance Computing on Vector Systems 2009*,
DOI 10.1007/978-3-642-03913-3_17

listed as one of the cooperative research and development items to be accelerated under the Global Nuclear Energy Partnership [2]. To fulfill the requirement to guarantee seismic safety, a series of experiments, which used a vibration table in the real world, was performed to rigorously measure the structural integrity of mechanical components. However, these experiments were performed only on independent sets of components because of the physical limitations of the experimental facilities employed. Hence, numerical simulations, which are capable of modeling entire nuclear power plants, are attracting increasing attention with the growth of computational technologies.

The analysis of huge assembled structure, which is aggregate of many components, differs greatly from the analysis of independent mechanical components because assembled structures are designed to function mechanically by each component interacting with other components in the structure. The center for computational science and e-systems, the Japan Atomic Energy Agency (CCSE/JAEA) has been conducting research on the 3D vibration simulator of nuclear power plants to establish a methodology for analyzing assembled structures and for accurately assessing the structural integrity of nuclear power plants under a seismic event to ensure a stable energy supply [3–5].

In this chapter, an overview of seismic simulation and a computation algorithm of large linear equations with domain decomposition method [6] is discussed. The balancing domain decomposition method [7] is a combination of primal substructuring approach [6, 8, 9], Neumann-Neumann preconditioning [10] and coarse grid correction [11] and its computation cost depends much on the number of subdomains. A prediction curve of parallel computation cost of balancing domain decomposition method is proposed and the optimal number of subdomains is estimated. Numerical validation of the optimal number of subdomains is conducted with a component of an actual nuclear power plant, High Temperature engineering Test Reactor (HTTR) [12], which is located at the O-arai research and development center of Japan Atomic Energy Agency. The component contains 48 millions degrees of freedom and successfully and efficiently analyzed on 128 and 512 processor elements on SGI Altix3700Bx2. The measured computation time with the optimal number of subdomains shows better computation performance than those with other numbers of subdomains.

2 Overview of Seismic Simulation

The Niigataken Chuetsu-oki (Kashiwazaki-Kariwa) earthquake in 2007 hit the Kashiwazaki-Kariwa nuclear power plants and it significantly exceeded the seismic load, which was taken into account at the design procedure of the plants. However, the safety installation worked in a safe manner and no significant damage was visibly observed by IAEA review team [13]. Although the nuclear power plants are generally designed to remain elastic during a forecasted seismic event [14], this fact indicates too conservative tolerance against the seismic load considered at the design stage and it is required to reassess the quantitative tolerance considering the ductile behaviour by computer simulation.

Structural analysis of nuclear power plants can be categorized in four fields, wide-range ground motion, ground-building interactions, dynamic interaction behaviour of mechanical components and detailed structural integrity evaluation of mechanical components as illustrated in Fig. 1. These four fields must interactively conducted to predict the structural integrity of nuclear power plants. The wide-range ground motion requires enough computation cost. The large scale simulation technology for wide-range ground motion has been successfully implemented by Lawrence Livermore National Laboratory [15] and we are working with them under the Global Nuclear Energy Partnership. Nonlinear simulation technologies for soil and buildings are rigorously investigated after the Northridge Earthquake (1994) and the Hyogoken Nanbu (Kobe) Earthquake (1995) and have greatly progressed. The current interests in this field are constitutive equation of concrete under cyclic loading, modelling of micro cracks, modelling of plastic hinges, interface description between mechanical components and concrete in reinforced concrete. The schematic view of groud-building interaction is illustrated in Fig. 2.

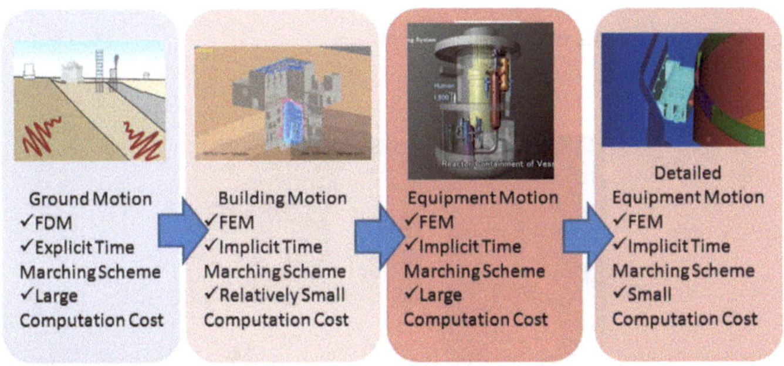

Fig. 1 Schematic flow of the seismic simulation of nuclear power plants

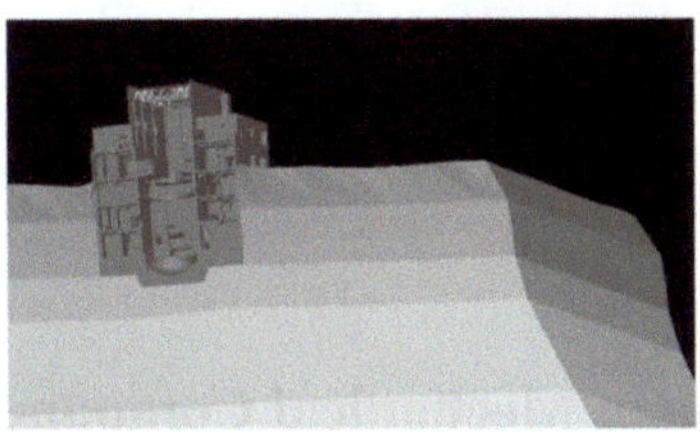

Fig. 2 Ground-building interactions

Other structural analyses are the ones concerning the seismic response of mechanical components and equipments inside and outside of a container vessel as illustrated in Fig. 3. These mechanical components are the reactor installed inside pressure vessel, cooling system components and pipes interconnecting these major components. A part of mechanical components inside a reactor are illustrated in Fig. 4. Because of the complexity and the huge number of mechanical components, these simulation requires full-scale 3D finite element analysis with enormous computation cost. The evaluation of the long term operated components and detailed simulation of ductile behaviour of each component are also a key issue for this kind of structural analysis.

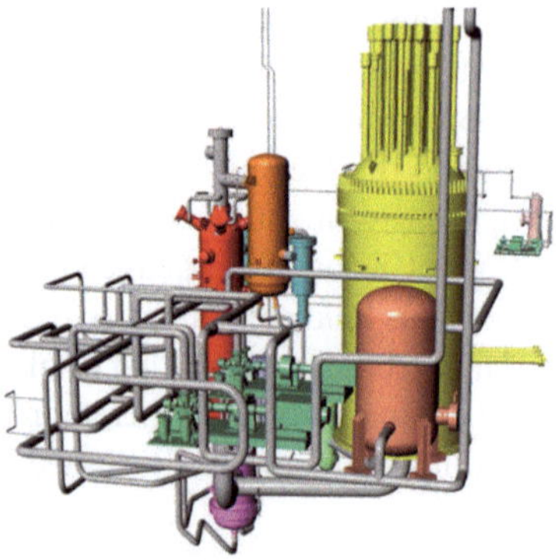

Fig. 3 Mechanical components in container vessel

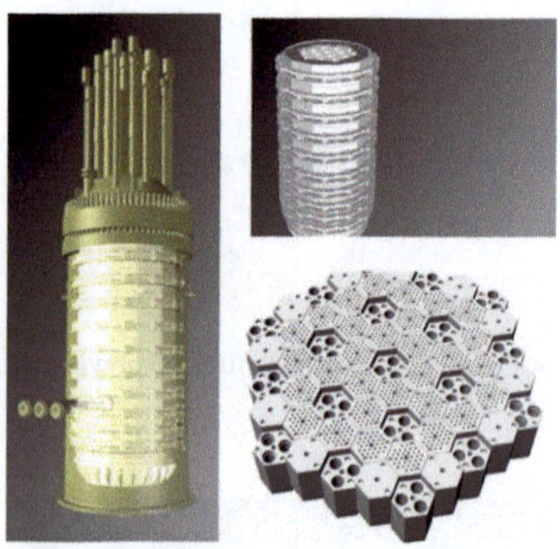

Fig. 4 Detailed inside components of reactor pressure vessel

3 Seismic Simulation of Mechanical Components

3.1 Governing Equations

The seismic simulation of mechanical components requires much computation cost because mechanical components interact with each other and full-scale 3D finite element analysis must be conducted to predict the real and detailed dynamic responses. The transient behaviour of an elastic body subjected to infinitesimal deformation is assumed in our current simulation. Therefore, the governing equation is given as follows,

$$\rho \ddot{\mathbf{u}} = \nabla \cdot \boldsymbol{\sigma} + \mathbf{f}, \tag{1}$$

where ρ, $\boldsymbol{\sigma}$, $\ddot{\mathbf{u}}$ and $\mathbf{f}$ represent density, Cauchy stress tensor, acceleration vector and body force, respectively. This governing question is discretized with finite element approximation and simulated with prescribed displacements and tractions. These essential and Neumann boundary conditions can be described as follows,

$$\mathbf{t} = \boldsymbol{\sigma} \cdot \mathbf{n} = \bar{\mathbf{t}}, \tag{2}$$

$$\mathbf{u} = \bar{\mathbf{u}}, \tag{3}$$

where $\mathbf{n}$ is unit vector normal to boundary, and $\bar{\mathbf{t}}$ and $\bar{\mathbf{u}}$ are prescribed traction and displacement vectors, respectively. The constitutive equation of an elastic body under infinitesimal deformation is given as follows,

$$\boldsymbol{\varepsilon} = \nabla_s \mathbf{u}, \tag{4}$$

$$\boldsymbol{\sigma} = \mathbf{C} : \boldsymbol{\varepsilon}, \tag{5}$$

where $\boldsymbol{\varepsilon}$ is the infinitesimal strain, ∇_s is the symmetric part of the gradient and $\mathbf{C}$ is a 4th order tensor relating the strain and Cauchy stress tensors. Through the Galerkin finite element discretization procedures, following equation in matrix form is obtained.

$$\mathbf{M}\ddot{\mathbf{u}} + \mathbf{K}\mathbf{u} = \mathbf{F}, \tag{6}$$

where $\mathbf{M}, \mathbf{K}, \mathbf{F}$ represent mass and stiffness matrices and a force vector. Equation (6) is advanced in time-domain by the Newmark's β method.

3.2 Balancing Domain Decomposition Method

To efficiently conduct a large scale structural component analysis, balancing domain decomposition method is employed. The balancing domain decomposition method is based on primal substructuring approach combined with Neumann-Neumann preconditioner and coarse grid correction in coarse space constructed with rigid body motion in each subdomain. The linear algebraic equations to be solved can be described as follows,

$$\mathbf{K}\mathbf{u} = \mathbf{f}, \tag{7}$$

where $\mathbf{K}$ is positive, definite and symmetric matrix, $\mathbf{u}$ is the solution vector and $\mathbf{f}$ is a given right hand side vector. In primal substructuring approach, the analysis domain, Ω, is decomposed into non-overlapping subdomains, Ω_i. In each subdomain, the subdomain boundary, the local unknown vector and local stiffness matrix can be denoted as Γ_i, $\mathbf{u_i}$, $\mathbf{K_i}$, respectively. The global stiffness matrix and local stiffness matrix can be converted with appropriate conversion matrix, $\mathbf{R_i}$, which is a (0,1) integer matrix as follows,

$$\mathbf{K} = \sum_{i=1}^{N} \mathbf{R_i}^T \mathbf{K_i} \mathbf{R_i}, \tag{8}$$

where N is the number of subdomains. Each element in local vector and matrix is classified as interface element, which is allocated on the boundary of subdomain, and interior element, which is not allocated on the boundary. By eliminating the interior elements of local stiffness matrices, the reduced systems of (7) is obtained.

$$\mathbf{S}\mathbf{u_B} = \mathbf{g}, \tag{9}$$

where $\mathbf{S}$ is the Schur complement of $\mathbf{K}$, $\mathbf{u_B}$ is the unknown vector on the interface of all the subdomains, and $\mathbf{g}$ is the corresponding given vector. This reduced system is called as interface problem and generally preconditioned conjugate gradient method is applied for its solution procedure. The preconditioned conjugate gradient method contains the following auxiliary problem,

$$\mathbf{M}\mathbf{z} = \mathbf{r}, \tag{10}$$

where $\mathbf{M}$ is a preconditioning matrix and $\mathbf{r}$ is a residual vector in each iterative step. The Neumann-Neumann preconditioner in balancing domain decomposition method is described as follows,

$$\mathbf{M_{N-N}}^{-1} = \sum_{i=1}^{N} \mathbf{R_{Bi}}^T \mathbf{D_i}^T \mathbf{S_i}^+ \mathbf{D_i} \mathbf{R_{Bi}}, \tag{11}$$

where $\mathbf{S_i}^+$ is a generalized inverse of $\mathbf{S_i}$, and $\mathbf{D_i}$ is a weight matrix. The preconditioning matrix for balancing domain decomposition method is introduced as follows,

$$\mathbf{M_{BDD}}^{-1} = \mathbf{R_c}\mathbf{S_c}^{-1}\mathbf{R_c}^T + \left(\mathbf{I} - \mathbf{R_c}\mathbf{S_c}^{-1}\mathbf{R_c}^T\mathbf{S}\right)\mathbf{M_{N-N}}^{-1}\left(\mathbf{I} - \mathbf{S}\mathbf{R_c}\mathbf{S_c}^{-1}\mathbf{R_c}^T\right), \tag{12}$$

where $\mathbf{S_c}$ and $\mathbf{R_c}$ are a coarse matrix and restriction matrix to construct coarse matrix. Hence, $\mathbf{S_c}$ is defined as follows,

$$\mathbf{S_c} = \mathbf{R_c}^T \mathbf{S} \mathbf{R_c} \tag{13}$$

The coarse problem is a Galerkin projection of original structural problem on a coarse space derived from rigid body motion of each subdomain. When the residual vector, $\mathbf{r}$ is balanced, (12) can be simplified as follows.

$$\mathbf{M_{BDD}}^{-1} = \left(\mathbf{I} - \mathbf{R_c}\mathbf{S_c}^{-1}\mathbf{R_c}^T\mathbf{S}\right)\mathbf{M_{N-N}}^{-1} \tag{14}$$

3.3 *Optimization of Number of Subdomains in Balancing Domain Decomposition Method*

3.3.1 Computation Cost for Each Iteration

The number of iterations has little dependency on the number of subdomains in balancing domain decomposition method. Hence, the computation cost of balancing domain decomposition method mainly dominated by the computation cost for each iteration of preconditioned conjugate gradient solver. The time consuming procedures of the method are a matrix-vector multiplication of Schur complement and given vector for conjugate gradient procedure, Neumann-Neumann preconditioning and coarse grid correction. Assuming these calculations are conducted with direct solution procedures, all these calculations contain forward elimination and backward substitution in each iteration of conjugate gradient solver and their computation cost is roughly estimated by using the size of matrices. Because the matrix-vector multiplication of Schur complement and given vector and Neumann-Neumann preconditioning are performed in each subdomains, the cost of forward elimination and backward substitution must be multiplied by the number of subdomains.

3.3.2 Prediction Curve of Total Computation Cost

The prediction curve of total non-dimensional computation cost, Q, is obtained by the computation cost for each iteration.

$$Q = N_{fine}^2 / N_{subdomain} + 12 N_{subdomain}^2, \tag{15}$$

where N_{fine} is the number of degrees of freedom in analysis domain and $N_{subdomain}$ is the number of subdomains. In this prediction of computation cost, the number of unknowns in the matrix-vector multiplication of Schur complement and given vector and Neumann-Neumann preconditioning is roughly assumed as $N_{fine}/N_{subdomain}$. On the other hand, the number of unknowns in coarse grid correction is $6N_{subdomain}$.

In case of parallel computation, computation cost for each processor can be assumed by using parallel speed up factors. The matrix-vector multiplication of Schur complement and given vector and Neumann-Neumann preconditioning are performed in each subdomain independently, hence the parallel efficiency is quite high and the parallel speed up could be close to the number of processors engaged in the computation when the number of processor is adequate. On the other hand, the parallel speed up of coarse grid correction is limited by a saturation point lower than the number of processors, because the size of coarse matrix is comparatively small and sparse matrix solver has poor parallel efficiency compared to the iterative one. Here, the parallel computation cost on each processor can be estimated as follows,

$$Q_{parallel} = \frac{N_{fine}^2 / N_{subdomain}}{S_{up1}} + \frac{12N_{subdomain}^2}{S_{up2}}, \tag{16}$$

where S_{up1} and S_{up2} denote the parallel speed up of local calculation in each subdomain and coarse grid correction.

The optimal number of subdomains can be estimated by minimizing the above total computation cost.

$$N_{optimalsubdomain} = \sqrt[3]{\frac{N_{fine}^2}{24} \cdot \frac{S_{up2}}{S_{up1}}} \tag{17}$$

4 Numerical Validation on a Parallel Computer

To evaluate acceleration of balancing domain decomposition method by using optimized number of subdomains, a real component of nuclear power plant was analyzed. The elastostatic analysis of secondary pressurized water cooler with 48 millions degrees of freedom was conducted on 128 and 512 processors on SGI Altix3700Bx2, which is a distributed shared memory architecture with Itanium2 processors. The number of nodes and elements were 16,053,390 and 9,183,679, respectively. Figure 5 shows the computed von Mises stress distribution under gravity load. The original analysis code was developed by the ADVENTURE project [16–18], which was one of the research projects in the "Computational Science and Engineering" field selected for the "Research for the Future (RFTF)" Program sponsored by the Japan Society for the Promotion of Science (JSPS). The prediction of the optimal number of subdomains is performed with parallel speed up factors described in Table 1. The predicted optimal numbers of subdomains on 128 and 512

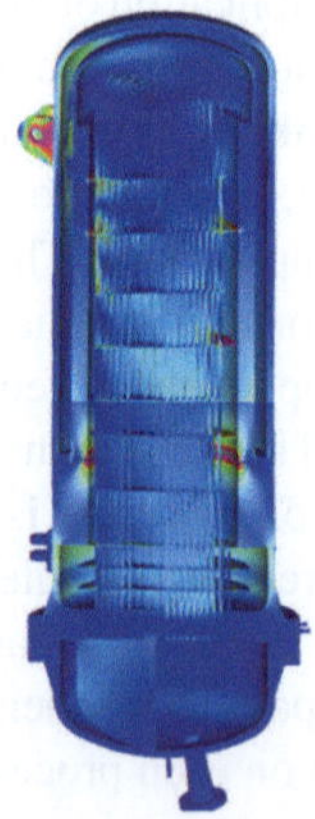

Fig. 5 Secondary pressurized water cooler subjected to gravity force

processor elements (PE) are approximately 18,200 and 11,500, respectively. The measured computation time of whole iterative procedure on 128 and 512PE are illustrated in Tables 2 and 3, respectively. In the results on 128PE, the minimum computation cost was achieved close to the predicted optimal number of subdomains. On the other hand, the minimum computation cost for whole iterative procedure was obtained with 22,528 subdomains as is illustrated in Table 4 because the number of iterations to achieve convergence was drastically reduced only in this case. However, the minimum computation time for one iterative step was obtained with 11,264 subdomains and it coincides with the predicted optimal number of subdomains.

Finally, the traditional scaled conjugate gradient method without substructuring required 0.51 million iterations and 13 hours on 512PE. The balancing domain decomposition method with optimal number of subdomains was approximately 30 times faster than the conventional conjugate gradient approach.

Table 1 Parallel speed up factors

Number of Processor Elements	Speed Up Factors	Coefficients
512	S_{up1}	512
512	S_{up2}	8
128	S_{up1}	128
128	S_{up2}	8

Table 2 Computation time on 128 processor elements

Number of Subdomains	Number of Iterations	Computation Time (sec.)
9,088	1,290	5,565
18,176	1,139	4,830
27,264	1,655	9,231
36,352	1,882	14,209

Table 3 Computation time on 512 processor elements

Number of Subdomains	Number of Iterations	Computation Time (sec.)
5,632	1,715	2,770
11,264	1,091	1,573
16,896	1,299	2,082
22,528	842	1,549

Table 4 Computation cost of each iteration on 512 processor elements

Number of Subdomains	Computation Cost of Each Iteration (sec.)
5,632	1.62
11,264	1.44
16,896	1.60
22,528	1.84

5 Concluding Remarks

An overview of seismic simulation of nuclear power plants and minimization methodology of computation cost in the balancing domain decomposition method were discussed. Although, the full-scale 3D finite element analysis of a nuclear power plant in its entirety requires huge computation cost, we try to reduce it with minimization of computation cost in the balancing domain decomposition method. The optimization of the number of subdomains was conducted on a parallel environment in this research. A mechanical component of actual nuclear power plant, HTTR, was analyzed with optimal number of subdomains and proposed optimization approach was validated.

References

1. NSC, MEXT and NISA: Nuclear Safety Research Forum 2008—Nuclear Safety Research and Seismic Safety on Nuclear Facilities (2008)
2. Ministry of Economy, Trade and Industry of Japan: United States—Japan Joint Nuclear Energy Action Plan (2007)
3. Nakajima, N., et al.: A large scale simulation for impact and blast loading issues. In: Proceedings of International Symposium on Structures under Earthquake, Impact, and Blast Loading 2008, pp. 125–128 (2008)
4. Nishida, A., et al.: Numerical simulation system "three-dimensional virtual plant vibration simulator" for nuclear plants by using assembled structural analysis. Trans. At. Energy Soc. Jpn. **6**, 376–382 (2007)
5. Tian, R., Yagawa, G.: Non-matching mesh gluing by meshless interpolation-an alternative to Lagrange multipliers. Int. J. Numer. Methods Eng. **71**, 473–503 (2007)
6. Toselli, A., Widlund, O.: Domain Decomposition Methods: Algorithms and Theory. Springer, Berlin (2004)
7. Mandel, J.: Balancing domain decomposition. Commun. Numer. Methods Eng. **9**, 233–241 (1993)
8. Yagawa, G., Shioya, R.: Parallel finite elements on a massively parallel computer with domain decomposition. Comput. Syst. Eng. **4**, 495–503 (1994)
9. Miyamura, T., et al.: Elastic-plastic analysis of nuclear structures with millions of DOF using the hierarchical domain decomposition method. Nucl. Eng. Des. **212**, 335–355 (2002)
10. Roeck, Y., Tallec, P.: Analysis and test of a local domain decomposition preconditioner. In: Fourth International Symposium on Domain Decomposition Methods for Partial Differential Equations, pp. 112–128 (1991)
11. Trottenberg, U., Oosterlee, C., Schuller, A.: Multigrid. Academic Press, San Diego (2000)
12. Shiozawa, S., et al.: Overview of HTTR design features. Nucl. Eng. Des. **233**, 11–22 (2004)
13. IAEA: Preliminary Findings and Lessons Learned from the 16 July 2007 Earthquake at Kashiwazaki-Kariwa NPP. IAEA report (2007)
14. Choi, H., Kim, J.: Energy-based seismic design of buckling-restrained braced frames using hysteretic energy spectrum. Eng. Struct. **28**, 304–311 (2006)
15. Appelo, D., Petersson, N.A.: A stable finite difference method for the elastic wave equation on complex geometries with free surfaces. Commun. Comput. Phys. **5**, 84–107 (2008)
16. Ogino, M., Shioya, R., Kanayama, H.: An inexact balancing preconditioner for large-scale structural. J. Comput. Sci. Technol. **2**, 150–161 (2008)
17. Yoshimura, S., et al.: Advanced general-purpose computational mechanics system for large-scale analysis and design. J. Comput. Appl. Math. **49**, 279–296 (2000)
18. ADVENTURE Project: url http://adventure.sys.t.u-tokyo.ac.jp

Understanding Electron Transport in Atomic Nanowires from Large-Scale Numerical Calculations

S. Wippermann, N. Koch, S. Blankenburg, U. Gerstmann, S. Sanna, E. Rauls, A. Hermann, W.G. Schmidt

Abstract Using the prototypical example of Si substrate-supported In nanowires we study the influence of adatoms (In, Pb, H, O) on the Landauer conductance on atomic-scale nanowires from first-principles calculations. Despite the increase of the total (and partially even local) density of states at the Fermi level due to the adsorption, all adatom species lower the nanowire conductance. Apart from hydrogen, which barely changes the transport properties, the conductance drop is pronounced, ranging from 17% for Pb to 38% for In. It is related to potential-well scattering and/or structural deformations of the nanowires.

1 Introduction

The atom-by-atom fabrication and modification of low-dimensional nanostructures is one of the impressive technological advances of recent years. It allows for the manipulation and positioning of single atoms or molecules in nanodevices such as artificial atomic-scale wires, see e.g., [1, 2]. Such quasi one-dimensional (1D) structures do not only show fascinating physical properties, they also have a large technological potential, e.g., as atomic-scale interconnects. Therefore there is an increasing interest in understanding and predicting the electronic properties of quasi-1D systems [3, 4]. Highly anisotropic surface superstructures have attracted considerable attention in this context.

The ordered array of In "nanowires" formed on Si(111) upon room temperature (RT) In monolayer deposition and subsequent annealing is probably the most intensively investigated system of this kind. By now, experiment and theory have established a structural model for its RT phase (see Fig. 1 for a top view) that is well

S. Wippermann, N. Koch, S. Blankenburg, U. Gerstmann, S. Sanna, E. Rauls, W.G. Schmidt
Lehrstuhl für Theoretische Physik, Universität Paderborn, 33095 Paderborn, Germany

A. Hermann
Institute of Fundamental Sciences, Massey University, Auckland, New Zealand

M. Resch et al. (eds.), *High Performance Computing on Vector Systems 2009*,
DOI 10.1007/978-3-642-03913-3_18 © Springer-Verlag Berlin Heidelberg 2010

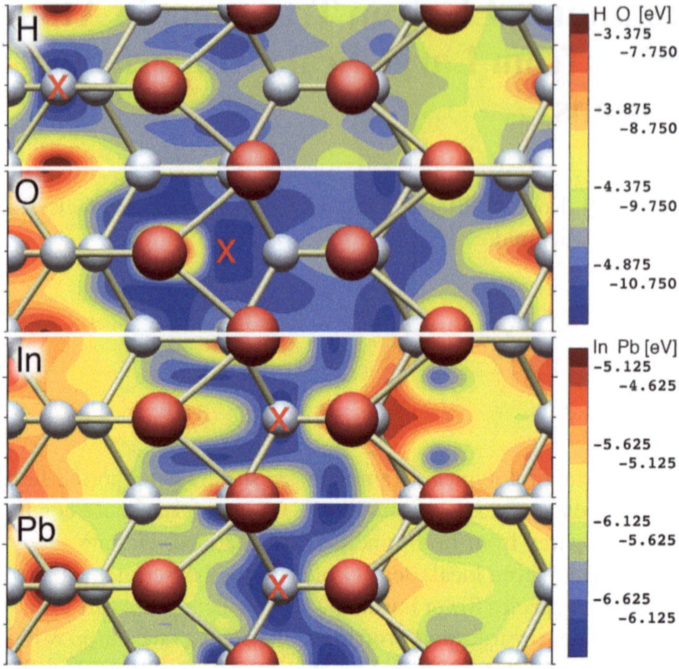

Fig. 1 Potential energy surface calculated for H, O, In, and Pb adatoms on the In/Si(111)(4×1) surface. Large and small balls indicate In and Si atoms, respectively. The minimum energy positions for the adsorbed species are marked

accepted [5–13]: Each nanowire consists of two zigzag chains of In atoms within a (4 × 1) surface periodicity. There are still many open questions, concerning, e.g., the temperature-induced (4 × 1) → (8 × 2) phase transition of the In overlayer structure [14–17], the modification of its structural and electronic properties upon adsorption of foreign as well as In adatoms [18–20], and the electron transport properties of ideal and defective In nanowires [21–23, 17]. It was found that the conductivity of the—at RT metallic [21, 22]—In nanowires is roughly halved by adsorbing additional In (0.1 ML) on top of the wires [23]. The mechanism that quenches the conductance, and the influence of other species on the wire resistance is not known, however, and is investigated here computationally, see [24].

2 Computational Method

Our calculations are based on density functional theory (DFT) within the local density approximation (LDA). DFT calculations have difficulty to describe accurately the subtle energetics of the temperature induced (4 × 1) → (8 × 2) nanowire phase transition [16, 17]. On the other hand, all measured properties of the RT (4 × 1) phase are well reproduced within DFT [8, 9, 11, 13]. Therefore, and because the potential energy surfaces (PES) of the adatoms considered here (see Fig. 1) turn out to be highly corrugated, we expect DFT calculations to be adequate for ideal and adatom perturbed In/Si(111)(1 × 4) surfaces.

Within the DFT calculations, the system of Kohn-Sham equations

$$\left\{-\frac{\hbar^2}{2m}\Delta + V_{ext}(\mathbf{r}) + \int \frac{n(\mathbf{r}')}{|\mathbf{r}-\mathbf{r}'|}d\mathbf{r}' + V_{xc}(\mathbf{r})\right\}\psi_{n\mathbf{k}}(\mathbf{r}) = \varepsilon_{n\mathbf{k}}\psi_{n\mathbf{k}} \quad (1)$$

$$n(\mathbf{r}) = \sum_{n,\mathbf{k}} f_{n\mathbf{k}}|\psi_{n\mathbf{k}}|^2 \quad (2)$$

is solved iteratively until self-consistency in the total electron density $n(\mathbf{r})$ is reached. In detail, we use a $8 \times 2 \times 1$ **k**-point mesh to sample the Brillouin zone and a plane-wave basis with an energy cutoff of 400 eV for the expansion of the wave functions $\psi_{n\mathbf{k}}(\mathbf{r})$.

For the DFT calculations, we ported the PWscf code [25] of the QUANTUM espresso package—frequently used in computational materials science—to the SX-8 vector machine. By cleaning up the code, e.g. removing all function calls within high dimensional loops, we obtain an efficient vectorization of these time-consuming loops and, by this, an excellent performance of the code. The computational benchmarks for a self-consistent DFT calculation for 300 atoms (a typical system size for our calculations) are: 35.4 GB memory, 83.57 hours cpu time, 99.62% vector optimization ratio and 14.431 GFLOPS.

Based on the electronic structure obtained within DFT, the Landauer conductance is calculated. Thereby the quantum conductance G representing the microscopic quantity that characterizes the transport properties of a conductor is given by

$$G(E) = \frac{2e^2}{h}T(E) \quad (3)$$

where $T(E)$ is the transmission function, i.e., the probability that an electron injected at one end of the conductor with energy E will be transmitted to the other end. In the Green's function formalism the transmission function T is expressed as

$$T = Tr(\Gamma_L G_C^r \Gamma_R G_C^a) \quad (4)$$

where $G_C^{r,a}$ are the retarded and advanced Green's functions of the conductor and $\Gamma_{L,R}$ are functions that describe the coupling of the conductor to the leads. The Green's functions G_C and the coupling functions Γ are explicitly given in dependence of the Hamiltonian matrix of the conductor and the self-energies that account for the coupling of the conductor to the leads. They can be expressed in terms of transfer and coupling matrices that—as well as the Hamiltonian matrix—may be expanded in a localized-orbital basis [26]. By choosing maximally localized Wannier functions [27], essentially an exact mapping of the electronic ground state onto a minimal basis is provided. We use the WanT implementation [28] for calculating the coherent electron transmittance. Thereby we exploit the fact that the In-related surface states close to the Fermi energy (E_F)F depend very little on the substrate [29] and study 1D model structures that contain only the In and nearest neighbor Si atoms with the remaining Si dangling bonds terminated with hydrogen. Test calculations for structures with one and two Si bilayers [30] confirm the validity of this approach. For the transport calculations a $16\times1\times1$ **k**-point mesh and an energy cutoff of 250 eV are found to lead to converged spectra. The memory requirements for the transport calculations are relatively modest, of the order of 16 GB. The CPU time requirements, however, are comparable to the DFT calculations discussed above. Therefore a parallelization of the calculations using up to 32 processors is done.

3 Results

We start by determining the PES for H, O, In, and Pb adsorbed on the nanowire array. These calculations are performed in a (4×3) unit cell with three bilayers of Si. The calculated energy landscapes are shown in Fig. 1. In and Pb prefer a position between neighboring In chains, O adsorbs threefold coordinated on top of the In chain and H bonds to a surface silicon atom. While the lateral positions of In and Pb adatoms are nearly equal, the adsorption geometry is not. The In adatom is located almost in-plane with the In chain atoms, whereas Pb prefers a pyramidal configuration, see Fig. 2. This affects also the structural relaxation of the In nanowire. Comparatively small deformations occur for Pb, but large atomic shifts are found for In deposition. This is reflected in the standard deviations of the In–In bond length distribution and the average atomic shifts compiled in Table 1.

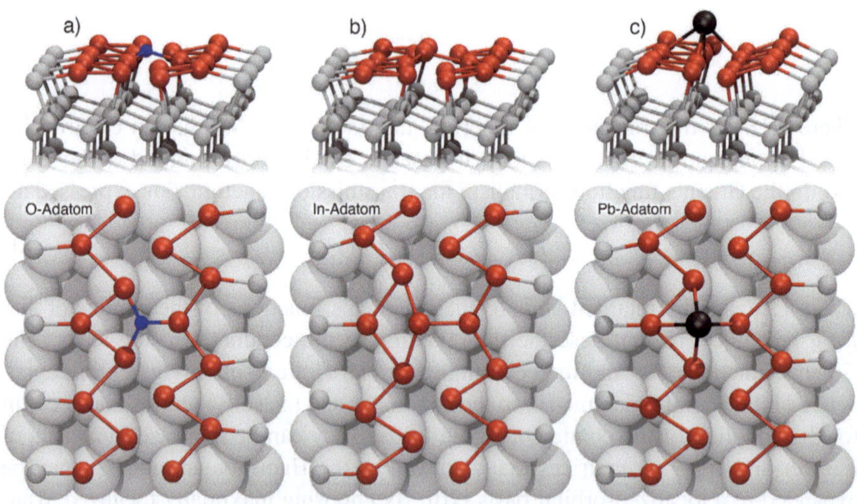

Fig. 2 Side and top views of (a) O, (b) In and (c) Pb adatoms adsorbed on the In/Si(111)(4×1) surface

Table 1 Standard deviation σ of the In–In bond length distribution and average shift $\bar{\Delta}$ of the In chain atoms for ideal and defect-modified nanowires. Average quantum conductance G in the energy interval ±0.05 eV around E_F, G' refers to the respective adatom structure without the adatom. The last column contains the electronic density of states at E_F

adatom	σ(Å)	$\bar{\Delta}$(Å)	$G(\frac{2e^2}{h})$	$G'(\frac{2e^2}{h})$	DOS (a.u.)
ideal	0.01	0.00	3.75	3.75	0.70
In	0.12	0.38	2.31	2.77	0.95
Pb	0.07	0.16	3.11	3.72	1.23
H	0.04	0.21	3.60	3.74	1.44
O	0.11	0.24	2.43	3.25	1.22

Starting from the relaxed adatom positions, the influence of the perturbation on the In nanowire conductance is calculated by using a lead-conductor-lead partitioning of the system. Here the In chain segment with the adatom forms the conductor (within a (8 × 8) wire segment) and the semi-infinite leads are modeled with ideal In nanowires. The calculated conductance spectra of the ideal and adatom perturbed In nanowires are shown in Fig. 3. The transmittance at E_F calculated for the unperturbed nanowire somewhat overestimates—by roughly a factor of two—the experimentally determined surface state conductance in the RT regime [21, 23, 22]. This can be attributed to the effect of the contacts and their scattering as well as to the thermal dissipative scattering due to phonons at finite temperature. These effects arise as well for perturbed In nanowires. However, the phonon scattering—neglected here—will modify the electron transport to different degrees [24].

Compared to the calculations for the ideal structure, a reduction of the quantum conductance at E_F by more than one third compared to the ideal chain is calculated for the case of In adatoms. While the experimental conditions are not sufficiently well defined to allow for a quantitative comparison, the calculated reduction is of the same order as measured [23]. Interestingly, apart from hydrogen which does not substantially modify the electron transport properties, the calculations predict distinct conductance drops as well for Pb and in particular for O, see Table 1. One might want to explain this finding with a reduced density of electronic states (DOS) at E_F. However, as shown in Fig. 3 and Table 1, (i) the DOS increases irrespective of the specific adatom deposited and (ii) the DOS of the perturbed nanowires shows at best a very weak correlation with the conductance. For example, the DOS at E_F is nearly equal for Pb and O. However, O is far more effective in reducing the conductance. Obviously, the DOS does not suffice to understand the trend of the conductance change.

This holds also for the local density of states (LDOS) at E_F. In all cases we find a distinct and adatom specific LDOS modification upon adsorption, see Fig. 4. For example, In causes a sharp local DOS increase close to the adatom that is accompanied by a DOS depletion at the next nearest neighbor distance. The adsorption of O leads to a DOS redistribution from the neighboring to the adatom decorated In chain. Even hydrogen—that does not affect the nanowire conductance and adsorbs on Si atoms rather than on the In chain—clearly affects the nanowire LDOS. Thus,

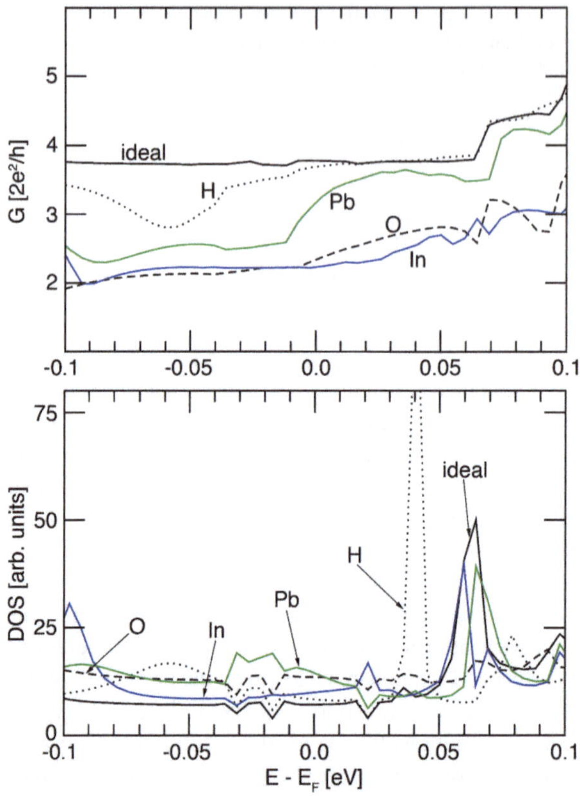

Fig. 3 Quantum conductance spectrum for electron transport along the wire direction (upper part) and total density of states (lower part) calculated for ideal and adatom-modified In/Si(111) structures

neither the total nor the local DOS seem very helpful for understanding the influence of adatoms on the nanowire conductance.

In the case of CO adsorption on substrate-supported Au chains, the drastic conductance drop could be traced to the deep potential well arising at the adsorption site [4]. In order to see if a similar mechanism acts here, we extract the local effective potential $V_{eff}(\mathbf{r}) = V_{ext}(\mathbf{r}) + \int \frac{n(\mathbf{r}')}{|\mathbf{r}-\mathbf{r}'|}d\mathbf{r}' + V_{xc}(\mathbf{r})$ in (1) from our DFT calculations. This potential is then averaged in a plane perpendicular to the nanowire direction chosen large enough to contain—within their covalent radii—the nanowire In atoms as well as the adatoms. As shown in Fig. 5, the systems studied here differ drastically with respect to the local potential. A very deep potential well is formed upon Pb adsorption, while additional In atoms barely change the potential along the wire direction. Thus it seems likely that the conductance modification upon Pb adsorption obeys a similar mechanism as proposed for CO adsorbed on Au chains [4]. Judging from Fig. 5, however, this mechanism cannot explain the conductance drop upon In deposition: In adatoms reduce the In nanowire transmittance even more than Pb,

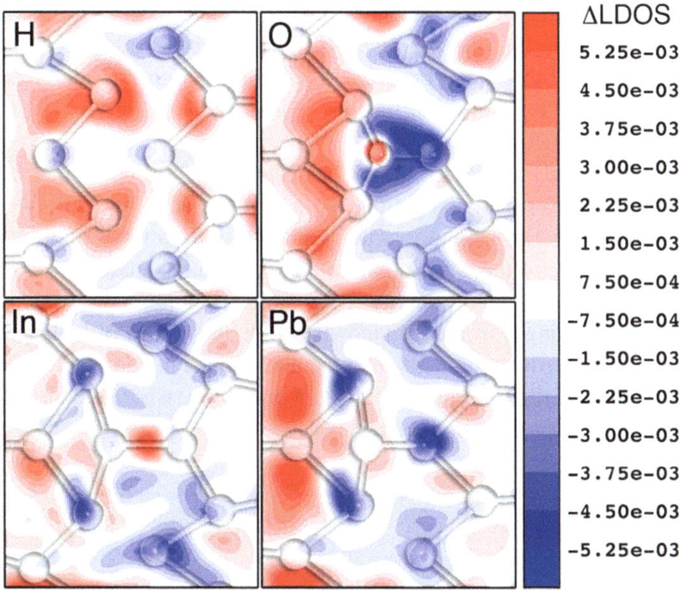

Fig. 4 Adsorption induced changes (in a.u.) of the local density of states at E_F projected on the plane of the Si(111) surface. Negative and positive values indicate local DOS depletion and accumulation with respect to the ideal In/Si(111) system, respectively

but do not give rise to large potential fluctuations. These considerations are corroborated by 1D model calculations, where we solve the time-dependent Schrödinger equation for the potentials shown in Fig. 5. The transmission of Fermi wave-vector electrons across the Pb potential well is reduced by 8%, while we obtain a reduction by only 3% in case of In.

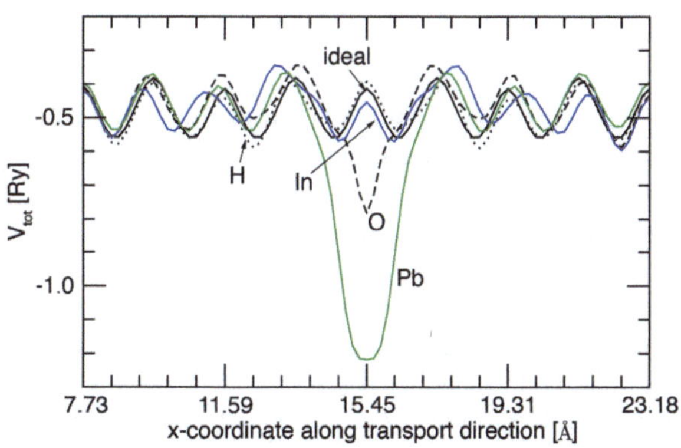

Fig. 5 Averaged (see text) effective potential along the wire direction calculated for ideal and adatom-modified In/Si(111) structures

In order to understand the conductance drop in the latter case, we come back to our initial observation that the adatoms deform the nanowire to different degrees, as can be seen from Table 1 and Figs. 2 and 4. While the smallest deformations are observed for H and Pb adsorption, In causes substantial strain. Oxygen represents an intermediate case.

The computational modeling allows for separating the impact of the adatom-induced structure deformation from the impact of the adatom itself: We perform transport calculations for nanowire structures that are deformed according to their relaxation in response to the adatom, but do *not* contain the adatom. The results are shown in Fig. 6 and compiled in Table 1. As can be seen here, the—comparatively small—geometry changes of the In nanowire upon adsorption of hydrogen or lead do not substantially reduce the wire conductance. This is in contrast to the stronger distortions caused by the adsorption of oxygen or indium, where moderate to strong conductance reductions are calculated.

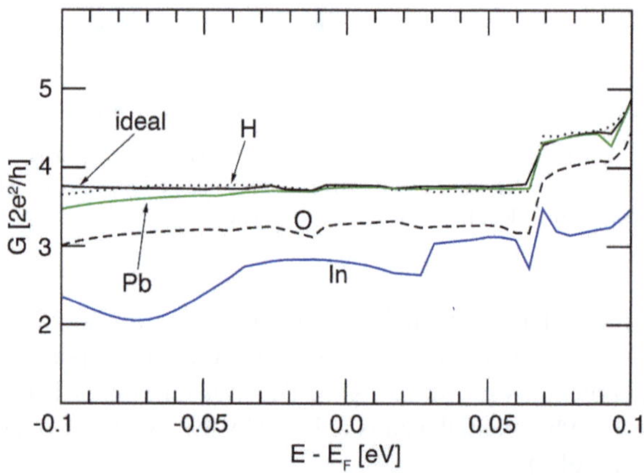

Fig. 6 Quantum conductance spectrum for electron transport along the wire direction calculated for ideal and adatom-modified In/Si(111) structures that do, however, not contain the adatom itself

This does now allow for a classification of the adatom induced conductance modifications. Pb adsorption does not substantially modify the nanowire geometry, but forms a deep potential well that effectively scatters the electrons and thus reduces the transmittance. No significant potential well forms for In. Here the conductance drop is related to the wire deformation. Both factors contribute in the case of oxygen. A moderate potential well is formed and on top of that the nanowire gets somewhat deformed. This results in a conductance drop of similar magnitude as calculated for In. Hydrogen, finally, does neither act as a potential well nor does it strain significantly the nanowire. Consequently, it has no substantial impact on the electron transport.

4 Summary

To conclude, the In nanowire array on Si(111)—intensively studied by many researchers—is a very interesting testbed also for conductance modification at the atomic scale: It is well accessible to *both* experiment and *first-principles* theory and thus helps to gain a deeper understanding of nanoscale electron transport. The *first-principles* calculations performed here for ideal and adatom deposited In nanowires predict an adatom-specific and in some cases very pronounced decrease of the wire conductivity upon adatom deposition. For In adatoms, where measurements exist, the reduction by more than one third agrees with the existing data. For oxygen deposition, the calculations predict a similar drop in conductance, whereas the impact of Pb atoms is slightly smaller. The adsorption of hydrogen does not substantially reduce the conductance. The nanowire conductance modification due to the adatoms can be related to different mechanisms: potential-well scattering (Pb), nanowire deformation (In) or a combination of both effects (O).

Acknowledgements Generous grants of computer time from the Höchstleistungs-Rechenzentrum Stuttgart (HLRS) and the Paderborn Center for Parallel Computing (PC^2) are gratefully acknowledged. We thank the Deutsche Forschungsgemeinschaft for financial support.

References

1. Nilius, N., Wallis, T.M., Ho, W.: Development of one-dimensional band structures in artificial gold chains. Science **297**, 1853 (2002)
2. Nazin, G.V., Qiu, X.H., Ho, W.: Visualisation and spectroscopy of a metal-molecule-metal bridge. Science **302**, 77 (2003)
3. Nilius, N., Wallis, T.M., Ho, W.: Localized molecular constraint on electron delocalization in a metallic chain. Phys. Rev. Lett. **90**, 186102 (2003)
4. Calzolari, A., Cavazzoni, C., Nardelli, M.B.: Electronic and transport properties of artificial gold chains. Phys. Rev. Lett. **93**, 096404 (2004)
5. Bunk, O., Falkenberg, G., Zeysing, J.H., Lottermoser, L., Johnson, R.L., Nielsen, M., Berg-Rasmussen, F., Baker, J., Feidenhans'l, R.: Structure determination of the indium-induced si(111)-(4 × 1) reconstruction by surface x-ray diffraction. Phys. Rev. B **59**, 12228 (1999)
6. Cho, J.H., Oh, D.H., Kim, K.S., Kleinman, L.: Weakly correlated one-dimensional indium chains on si(111). Phys. Rev. B **64**, 235302 (2001)
7. Nakamura, J., Watanabe, S., Aono, M.: Anisotropic electronic structure of the si(111)-(4 × 1) in surface. Phys. Rev. B **63**, 193307 (2001)
8. Miwa, R.H., Srivastava, G.P.: Atomic geometry, electronic structure and image state for the si(111)-in(4 × 1) nanowire. Surf. Sci. **473**, 123 (2001)
9. Wang, S., Lu, W., Schmidt, W.G., Bernholc, J.: Nanowire-induced optical anisotropy of the si(111)-in surface. Phys. Rev. B **68**, 035329 (2003)
10. Fleischer, K., Chandola, S., Esser, N., Richter, W., McGilp, J.F.: Phonon and polarized reflectance spectra from si(111)-(4 × 1)in: Evidence for a charge-density-wave driven phase transition. Phys. Rev. B **67**, 235318 (2003)
11. Cho, J.H., Lee, J.Y., Kleinman, L.: Electronic structure of one-dimensional indium chains on si(111). Phys. Rev. B **71**, 081310(R) (2005)

12. Tsay, S.F.: Atomic and electronic structure of the (4A-1) and (8A-2) in/si(111) surfaces. Phys. Rev. B **71**, 035207 (2005)
13. Lopez-Lozano, X., Krivosheeva, A., Stekolnikov, A.A., Meza-Montes, L., Noguez, C., Furthmüller, J., Bechstedt, F.: Reconstruction of quasi-one-dimensional in/si(111) systems: Charge- and spin-density waves versus bonding. Phys. Rev. B **73**, 035430 (2006)
14. Park, S.J., Yeom, H.W., Ahn, J.R., Lyo, I.W.: Atomic-scale phase coexistence and fluctuation at the quasi-one-dimensional metal-insulator transition. Phys. Rev. Lett. **95**, 126102 (2005)
15. Guo, J., Lee, G., Plummer, E.W.: Intertwined electronic and structural phase transitions in the in/si(111) interface. Phys. Rev. Lett. **95**, 046102 (2005)
16. Gonzalez, C., Flores, F., Ortega, J.: Soft phonon, dynamical fluctuations, and a reversible phase transition: Indium chains on silicon. Phys. Rev. Lett. **96**, 136101 (2006)
17. Stekolnikov, A.A., Seino, K., Bechstedt, F., Wippermann, S., Schmidt, W.G., Calzolari, A., Buongiorno Nardelli, M.: Hexagon versus trimer formation in nanowires on si(111): Energetics and quantum conductance. Phys. Rev. Lett. **98**, 026105 (2007)
18. Lee, G., Yu, S.Y., Kim, H., Koo, J.Y.: Defect-induced perturbation on si(111)4A-1-in: Period-doubling modulation and its origin. Phys. Rev. B **70**, 121304(R) (2004)
19. Lee, S.S., Ahn, J.R., Kim, N.D., Min, J.H., Hwang, C.G., Chung, J.W., Yeom, H.W., Ryjkov, S.V., Hasegawa, S.: Adsorbate-induced pinning of a charge-density wave in a quasi-1d metallic chains: Na on the in/si(111)-(4A-1) surface. Phys. Rev. Lett. **88**, 196401 (2002)
20. Ryjkov, S.V., Nagao, T., Lifshits, V.G., Hasegawa, S.: Phase transition and stability of si(111)-8×2-in surface phase at low temperatures. Surf. Sci. **488**, 15 (2001)
21. Uchihashi, T., Ramsperger, U.: Electron conduction through quasi-one-dimensional indium wires on silicon. Appl. Phys. Lett. **80**, 4169 (2002)
22. Kanagawa, T., Hobara, R., Matsuda, I., Tanikawa, T., Natori, A., Hasegawa, S.: Anisotropy in conductance of a quasi-one-dimensional metallic surface state measured by a square micro-four-point probe method. Phys. Rev. Lett. **91**, 036805 (2003)
23. Tanikawa, T., Matsuda, I., Kanagawa, T., Hasegawa, S.: Surface-state electrical conductivity at a metal-insulator transition on silicon. Phys. Rev. Lett. **93**, 016801 (2004)
24. Wippermann, S., Koch, N., Schmidt, W.G.: Adatom-induced conductance modification of in nanowires: Potential-well scattering and structural effects. Phys. Rev. Lett. **100**, 106802 (2008)
25. http://www.pwscf.org/
26. Datta, S.: Electronic Transport in Mesoscopic Systems. Cambridge University Press, Cambridge (1995)
27. Lee, Y.S., Nardelli, M.B., Marzari, N.: Band structure and quantum conductance of nanostructures from maximally localized Wannier functions: The case of functionalized carbon nanotubes. Phys. Rev. Lett. **95**, 076804 (2005)
28. Calzolari, A., Marzari, N., Souza, I., Nardelli, M.B.: Ab initio transport properties of nanostructures from maximally localized Wannier functions. Phys. Rev. B **69**, 035108 (2004)
29. Riikonen, S., Ayuela, A., Sanchez-Portal, D.: Metal-insulator transition in the in/si(111) surface. Surf. Sci. **600**, 3821 (2006)
30. Wippermann, S., Schmidt, W.G., Calzolari, A., Nardelli, M.B., Stekolnikov, A.A., Seino, K., Bechstedt, F.: Quantum conductance of in nanowires on si(111) from *first principles* calculations. Surf. Sci. **601**, 4045 (2007)

Multi-scale Simulations for Laser Plasma Physics

Hideo Nagatomo, Tomoyuki Johzaki, Kunioki Mima, Hitoshi Sakagami

Abstract The fast ignition scheme is one of the most fascinating and feasible ignition schemes for the inertial fusion energy. At ILE Osaka University, FIREX (Fast Ignition Realization Experiment) project is in progress. Implosion experiments of the cryogenic target are scheduled in near future. There are two key issues for the fast ignition. One is controlling the implosion dynamics to form high density core plasma in non-spherical implosion, and the other is heating the core plasma efficiently by the short pulse high intense laser. The time and space scale in the fast ignition scheme vary widely from initial laser irradiation to solid target, to relativistic laser plasma interaction and final fusion burning. The numerical simulation plays an important role in demonstrating the performance of the fast ignition, designing the targets, and optimizing laser pulse shapes for the scheme. These all the physics are desired to be self-consistently described. In order to study these physics of FI, we have developed "Fast Ignition Integrated Interconnecting code" (FI3), which consists of collective Particle-in-Cell (PIC) code (FISCOF1D/2D), Relativistic Fokker-Planck with hydro code (FIBMET), and two-dimensional Arbitrary-Lagrangian-Eulerian (ALE) radiation hydrodynamics code (PINOCO). Those codes are sophisticated in each suitable plasma parameters, and boundaries conditions and initial conditions for them are imported/exported to each other by way of DCCP, a simple and compact communication tool which enable these codes to communicate each other under executing different machines. The feature of the FI3 code, and a numerical result of integrated simulation are shown. This simulation system can be applied for particle acceleration and other applications.

Hideo Nagatomo, Tomoyuki Johzaki, Kunioki Mima
Institute of Laser Engineering, Osaka University, Osaka, Japan

Hitoshi Sakagami
National Institute for Fusion Science, Toki, Japan

M. Resch et al. (eds.), *High Performance Computing on Vector Systems 2009*,
DOI 10.1007/978-3-642-03913-3_19

1 Introduction

Fast ignition is an attractive scheme of laser fusion [1], where a fuel shell is compressed by implosion laser (pulse duration; 1–20 ns) beams to be more than 1000 times of the solid density. At the maximum compression time, it is heated by a ultra-intense short pulse laser (pulse duration; 1–10 ps) to create a hot spot at the edge of the compressed fuel. In the other scheme, central-hot-spot ignition scheme, a highly uniform laser irradiation and strict power balance of multi-beam laser system are required to form a hot igniting spot at the center in the compressed fuel. In the fast ignition, such a requirement is relaxed, and all required is to achieve the high-density compression (Fig. 1).

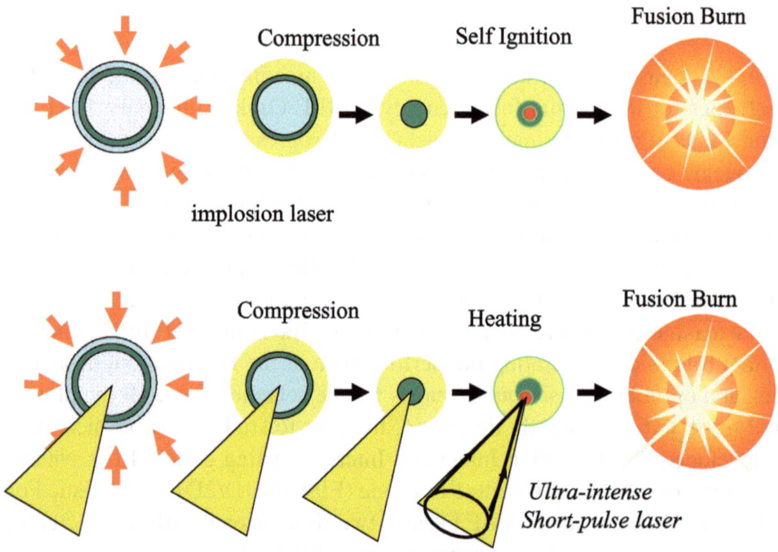

Fig. 1 Central-hot-spot ignition (above) and fast ignition scheme (below)

The integrated experiment for fast ignition research at Osaka University has been progressing since 2001. At the first stage of the experiments, the neutron yield increased from 104 without heating to 107, when ultra-intense laser light produced by a Peta-Watt laser (400 J/0.6 ps) was irradiated into compressed CD core plasma. In the result, the core plasma temperature increased by 500 eV, and it indicated that the energy coupling efficiency from heating laser to core plasma was 20–25% [2] In the fall of 2008, the construction of a new heating laser, LFEX (Laser for Fusion Experiment) was completed. LFEX has the highest performance in ultra-short pulse lasers, which has the capability of 10 kJ/10 ps at a wavelength of 1.06 μm. Using the LFEX laser, FIREX-I (Fast Ignition Realization Experiment) project has started [3, 4]. The final goal of FIREX-I is achieving the heated core plasma of 5 keV. After FIREX-I, second stage is planned to demonstrate the burning.

The numerical simulation plays an important role in demonstrating the performance of the fast ignition, designing the targets, and optimizing laser pulse shapes for the scheme. There are two key issues in the fast ignition. One is controlling the implosion dynamics to form high density core plasma in non-spherical implosion, and the other is heating the core plasma efficiently by the short pulse laser. The time and space scale in the fast ignition scheme vary widely from initial laser irradiation to solid target, to relativistic laser plasma interaction and final fusion burning. All the physical processes are desired to be self-consistently described in numerical calculation. However, it is a formidable task to simulate relativistic laser plasma interaction and radiation hydrodynamics in a single computational code, without any numerical dissipation, special assumption or conditional treatment. Therefore, multi-scale simulation is important in this task.

Recently, we have developed "Fast Ignition Integrated Interconnecting code" (FI3) [5] which consists of collective Particle-in-Cell (PIC) code (FISCOF1: Fast Ignition Simulation COde with collective and Flexible particles), Relativistic Fokker-Planck with hydro code (RFP-hydro) code [6], and two-dimensional Arbitrary-Lagrangian-Eulerian (ALE) radiation hydrodynamics code (PINOCO: Precision Integrated implosion Numerical Observation COde) [7]. Those codes are sophisticated in each suitable plasma parameters, and boundaries and initial conditions for them are imported/exported to each other by way of DCCP [8], a simple and compact communication tool which enable these code to communicate each others in different machines. In this paper, we will present the feature of FI3 code, and individual codes which consists of FI3. In each section, numerical methods and the latest numerical results of these codes which are related to fast ignition are introduced briefly. Finally, fully integrated simulation of fast ignition is presented.

2 Numerical Methods

Figure 2 shows the image of plasma density of fast ignition implosion and computational code which can solve those plasma parameters. These codes are best optimized for each plasma parameter regions to avoid undesired numerical dissipations or unwilling huge computing time. At first, cone-guided implosion dynamics is calculated by PINOCO because radiation hydrodynamics is dominant in implosion process. Near the maximum density timing, just before the irradiation of heating laser, the mass density, temperatures, and other profiles calculated by PINOCO are exported to both collective PIC and RFP-hydro code for their initial and boundary conditions.

The relativistic laser plasma interaction inside the cone target is simulated by collective PIC code, which exports the time-dependent energy distribution of fast electron to REP-hydro code. The fast electrons calculated by the FISCOF1 or 2 are exported to the RFP-hydro code. Therefore, the core heating process is simulated using both physical profiles of imploded core plasma and fast electron as the boundary conditions. The abstract of the profile data flows are illustrated in Fig. 2.

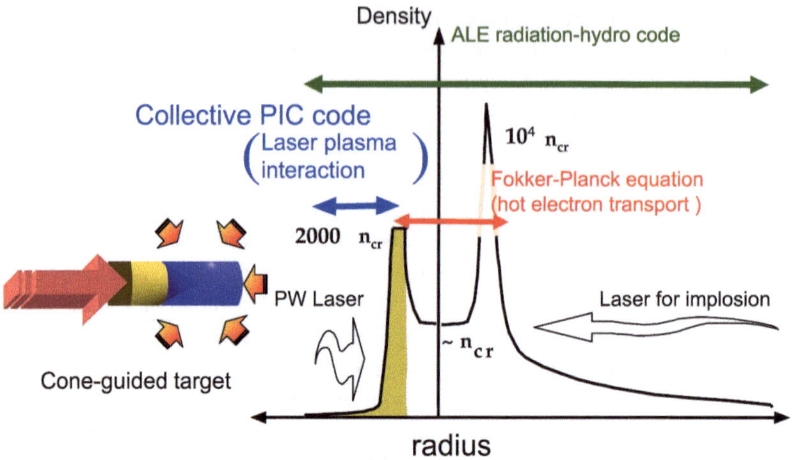

Fig. 2 Plasma density of fast ignition implosion and computational code which can solve those plasma parameters

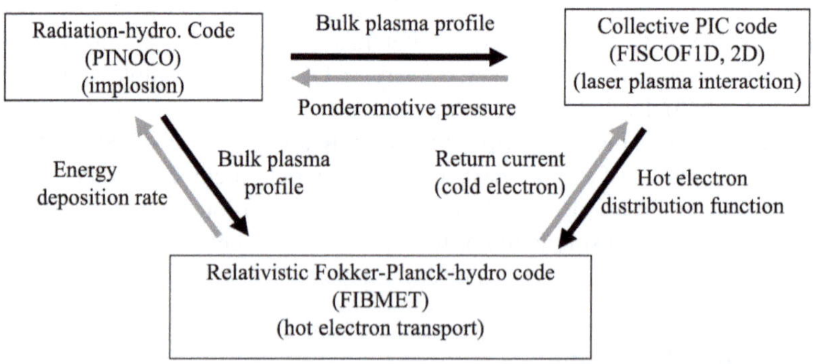

Fig. 3 Data flow in FI3 system. Black arrows are already executable data flows, and gray arrows are next plan to be considered

3 Radiation Hydrodynamics Code (PINOCO)

In PINOCO code, mass, momentum, electron energy, ion energy, equation of states, laser ray-trace, laser absorption, radiation transport, surface tracing and other related equations are solved simultaneously [5, 7]. In the most of the integrated implosion codes except PINOCO, hydrodynamic equations are solved by Lagrangian-based ALE method. But they are affected by numerical viscosity easily at rezoning/remapping process. Therefore, we have extended CIP (Cubic Interpreted Polynomial) method [9] into ALE type CIP method, so-call ALE-CIP. This modification has enabled the calculation of large dynamic range of the implosion. Originally, CIP has some characteristics of Lagrangian method, although the fundamental formulas are done for Eulerian coordinates. This CIP method is also employed to track the interface between the different materials clearly also. This tracking system is very useful when multi-material target structures must be considered. The equation of state is based on quotidian equation of state (QEOS) [10]. In the energy equations, flux-limited Spitzer-Harm type thermal transport model is solved using the implicit 9-point differencing for the diffusion equation with ILUBCG (Incomplete LU Biconjugate Gradient Conjugate) method. For the laser ray-tracing, a simple 1-D ray-tracing method is applied. The radiation transport solver was newly installed. Here, multi-group flux-limited diffusion-type equations are solved with ILUBCG implicit method. In the calculation of opacity and emissivity, LTE (Local Thermal Equilibrium) and CRE (Collisional Radiative Equilibrium) models are prepared for table lookup. Even though we can move the grid points as Lagrangian way in PINOCO simulation, so-called sliding-mesh in which the high resolution region is sliding along the mass center of the target is used in all these simulations. We have to remark that the advantage of the sliding-mesh is not only simple rezoning rule but also better convergence of the iteration method in solving the diffusion equations. Only an implosion code which is based on a high-order scheme such as PINOCO can simulate the problem with the sliding-mesh. In general, hydrodynamics code is well-suited to vector computers, which have high performance in memory access. In the same way, vector computers are appropriate for radiation hydrodynamics simulations.

4 Collective PIC Code (FISCOF1D and 2D)

Usually dynamics of interaction between electrons, ions, and electromagnetic field are solved in PIC method. In our study, for the evaluation of fast electron generation due to relativistic laser-plasma interactions (LPI), we use the 1D and 2D collective PIC code (FSCOF1, 2) [11, 12], where collective particles are used to represent many normal particles and then total number of particles and computations are drastically reduced. Even though, the FISCOF1D/2D code enables us to treat a wide range in space and high density region, the exact condition cannot be considered for the limitation of computer resources. Therefore, we simulated the small region of relativistic laser plasma interaction and this fast electron bean profile was extended to the realistic scale size in FI3. This assumption was temporal and will be eliminated after developing accelerated fully parallelized two-dimension Collective PIC code is developed in near future. In PIC simulations, large memory and large numbers of CPUs are required. Therefore, these simulations are executed in massive parallel computers.

5 Relativistic Fokker-Planck Hydrodynamic Code (FIBMET)

A 2-D Relativistic Fokker-Planck (RFP) code, FIBMET [6] has been developed for analysis of the fast electron transport and energy deposition processes in dense core plasma, which was coupled with an Eulerian hydrodynamic code to examine core-heating properties. In this code, cold bulk electrons and ions are treated by a 1-fluid and 2-temperature hydro model, and the fast electrons generated by the ignition laser-plasma interactions are treated by the RFP model. In the coupled RFP-hydro code includes magnetic field generated by fast electron current, gradient of plasma resistivity and pressure gradient. Initial and boundary condition can be imported from PINOCO and FISCOF1D/2D as data flows which are shown in Fig. 3. Numerical example of FIBMET is written in the latter section "Full integrated Simulation of Fast Ignition".

6 Distributed Computing Collaboration Protocol (DCCP)

We have three set of codes which have quite different properties in FI3 project. PINOCO and FIBMET codes are optimized for a vector parallel machine, and FISCOF1D/2D is optimized for a massive scalar parallel machine. Therefore, it is preferable to be combined with each other which use different machines by way of simple and easy communication tool for computational scientists. Since communication in our project is not complex and very straightforward, we design a special lightweight communication protocol, Distributed Computing Collaboration Protocol (DCCP), to transfer data between codes [8].

DCCP is implemented with two kinds of daemon programs and simulation code itself. One of daemon programs is called Communicator that actually transfers data instead of the code, and the other is called Arbitrator that manages communication between Communicators. The code does not send data directly to another code, but only asks the local Communicator, which is running in the background at his site, to transfer his data. The sender code, therefore, does not have to know details of the receiver code, such as IP address of the receiver computer. And then, the sender side Communicator inquires of the Arbitrator where is the remote Communicator that is handling the receiver code, and forwards the requested data to the appropriate Communicator via the Internet. The receiver side Communicator stores the data to storage, and is waiting for a demand. Finally the receiver code requires the data from the receiver side Communicator, and then the Communicator sends the data to the code and communication between two codes has just finished. If the receiver code does not run yet when the sender code sends data to the local Communicator, the Arbitrator orders it to save the data to storage for a future usage. Once the receiver code is invoked, the Arbitrator directs the sender side Communicator to restore and transfer data to the receiver side Communicator, and now the receiver code can get the desired data. Thus the sender code does not also have to take care of a situation whether the receiver code is running or not. Furthermore, if a broadband dedicated line is available between the Communicators, the Arbitrator tells both Communicators to use that dedicated line instead of the Internet and high-speed communication will be performed even both codes do not know about details of network configuration. TCP/IP based lightweight protocol does not transfer data directly to another code manage information of codes Arbitrator.

7 Fully Integrated Simulation of Fast Ignition

In this section, an example of fully integrated simulation using FI3 system is presented [13], where vector machine SX-6 and cluster machine are connected by DCCP. This simulation conditions correspond to the GXII experiment at ILE Osaka Univ. [1]. The detail profile of mass density, temperatures, and other conditions of imploded core plasma is imported from PINOCO simulation. The profiles of fast electrons generated in the laser plasma interaction were evaluated with FISCOF1D that includes both ion and electron motions but not the collision process. The Au cone tip was modelled by the 10 μm-thickness plasma. The 60 μm-thickness imploded plasma was put behind the cone tip. The simulations were carried out by assuming that the electron density in the rear of cone tip is ne, rear = 100 nc, 10 nc and 2 nc. The forward-directed fast electrons were observed behind the cone tip, and they are exported to FIBMET.

Using the imploded-core profile of PINOCO at $t = 1.97$ ns, and the time-dependent momentum profiles of relativistic electron beam obtained by FISCOF1D simulation in the cases of ne, rear = 2 nc, 10 nc and 100 nc, we evaluated the core heating properties using FIBMET. The REB source was injected behind the cone tip by assuming the super Gaussian profile with 30 μm width. Figure 4 shows the temporal profiles of bulk-electron and ion temperatures averaged over the core region ($\rho > 10$ g/cc). Compared with the results between 100 nc and 10 nc cases, the temporal evolution of core temperature is different because of the difference in intensity profiles of REBs. However, there are little differences in $\eta L \rightarrow e$ and the spectrum, so that the core temperatures reach the same value (0.45 keV) at the end

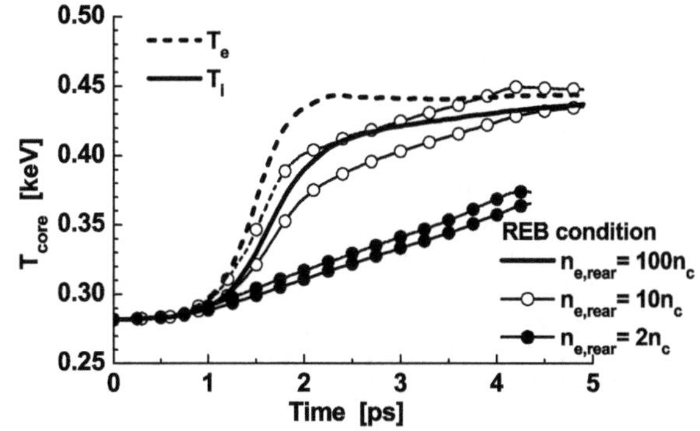

Fig. 4 Temporal profiles of bulk-electron and ion temperatures averaged over the dense core region ($\rho > 10$ g/cc) obtained for the three different REBconditions (ne, rear = 2, 10 and 100 nc)

of core heating. The core size (=0.14 g/cm^2) is smaller than the range of MeV electron (e.g., 0.6 g/cm^2 for 1 MeV electron), so that the most of fast electrons penetrate the core. The energy coupling from the REB to the core is 22% in both cases and then the coupling from the heating laser to the core is only 4.6%. In the case of ne, rear = 2 nc, the beam intensity and the $\eta L \rightarrow e$ are small compared with the other cases, so that the resultant core temperature is about 0.35 keV. This result does not exactly agree with the experiment [1] where core plasma was heated to 0.6 keV. We recognize some important physics are still missing, and it will be improved with the further developments of computational codes, and increments of computational resources.

8 Summary

Experiments, theoretical and computational researches for the fast ignition scheme have been implemented at ILE Osaka University. In order to simulate the whole process of the complex multi scale plasma physics in the fast ignition, we have developed FI3 system where three different computational codes are connected to each others. In this paper, we have introduced these latest numerical methods and results for the fast ignition. Multi-scale simulation is very important when realistic scale simulations are required. This code system can be applicable not only to the inertial fusion energy but also to the other applications, for example, particle acceleration, medical treatment, and laser nuclei physics.

Acknowledgements These simulations were executed at Cyber Media Center, Osaka University, and ILE Osaka University. The authors would like to appreciate the technical staffs at CMC and ILE Osaka University.

References

1. Tabak, M., et al.: Phys. Plasmas **1**, 1626 (1994)
2. Kodama, R., et al.: Nature **418**, 933 (2002)
3. Izawa, Y., et al.: IAEA-CN-116/OV/3-2 (2004)
4. Myanaga, N., et al.: J. Phys. IV Frances. EDP Sci. **133**, 81–87 (2006)
5. Nagatomo, H., et al.: IAEA-CN-116/IFP/07-29 (2004)
6. Johzaki, T., et al.: J. Plasma Fusion Res., Ser. **6**, 341 (2004)
7. Nagatomo, H., et al.: Phys. Plasmas **14**, 056303 (2007)
8. Sakaguchi, T., et al.: Parallel and Distributed Computing: Applications and Technologies. LNCS, vol. 3320, pp. 90–93. Springer, Heidelberg (2004)
9. Yabe, T., et al.: J. Comput. Phys. **169**, 556–593 (2001)
10. More, R.M., et al.: Phys. Fluids **31**(10), 3059–3078 (1988)
11. Sakagami, H.: Proceedings of IFSA 2001, pp. 434–437. Elsevier, Paris (2003)
12. Nakamura, T., et al.: J. Phys. IV Fr. **133**, 405–408 (2006)
13. Johzaki, T., et al.: J. Phys. IV Fr. **133**, 385–489 (2006)

MIX
Papier aus verantwortungsvollen Quellen
Paper from responsible sources
FSC® C105338

If you have any concerns about our products,
you can contact us on
ProductSafety@springernature.com

In case Publisher is established outside the EU,
the EU authorized representative is:
Springer Nature Customer Service Center GmbH
Europaplatz 3, 69115 Heidelberg, Germany

Printed by Libri Plureos GmbH
in Hamburg, Germany